MILK

the vital force

MILK
the vital force

POSTERS Presented at the
XXII International Dairy Congress
The Hague, September 29 - October 3, 1986

Edited by

International Dairy Federation

D. Reidel Publishing Company

A MEMBER OF THE KLUWER ACADEMIC PUBLISHERS GROUP

Dordrecht / Boston / Lancaster / Tokyo

Library of Congress Cataloging in Publication Data

CIP

International Dairy Congress (22nd : 1986 : The Hague, Netherlands)
Milk – the vital force.

1. Dairying—Congresses. 2. Milk—Congresses. I. International Dairy Federation. II. Title.
SF223.I6 1986 637'.1 86–17836
ISBN-13: 978-90-277-2330-7 e-ISBN-13: 978-94-009-3733-8
DOI: 10.1007/978-94-009-3733-8

Published by D. Reidel Publishing Company
P.O. Box 17, 3300 AA Dordrecht, Holland

Sold and distributed in the U.S.A. and Canada
by Kluwer Academic Publishers,
101 Philip Drive, Assinippi Park, Norwell, MA 02061, U.S.A.

In all other countries, sold and distributed
by Kluwer Academic Publishers Group,
P.O. Box 322, 3300 AH Dordrecht, Holland

FOREWORD

The 22nd International Dairy Congress was the first to offer to participants the opportunity to present their work in the form of a poster, the object being to include more contributions in the programme than there was time available for oral presentations. The posters were also intended to replace, up to a point, the brief communications that were received for, and published at, previous international dairy congresses but which had been criticized on various grounds.

The Organizing Committee's criteria for the acceptance of a poster included the following :

- the subject matter should not have appeared in print prior to the Congress
- the poster should contain useful and mainly new information
- the poster should have an adequate technical or scientific basis

Authors whose posters were accepted were invited to display their material at the Congress and be present to explain it to interested participants. The texts presented are also published in this volume.

The International Dairy Federation is interested in the views of Congress participants and of readers of this volume as to the relative value of posters vs brief communications. Any comments will be kept for consideration when future International Congresses are planned : they should be sent to IDF Secretariat at the address below.

IDF General Secretariat
41, square Vergote
B-1040 Brussels

September 1986

C O N T E N T S

M O N D A Y

T U E S D A Y

WEDNESDAY

T H U R S D A Y

EFFECTIVENESS OF DIFFERENT MODELS OF MILK COLLECTION

J. Kwiatkowski, M. Jurczak, M. Zajac
Institute of Cattle Breeding and Milk Dairy Science
Warsaw Agricultural University and The Central Union of Dairy Cooperatives, 05-840 Brwinow, Poland

With the aim of developing improved systems of milk production, collection and transport, five different models have been implemented and their effectiveness was tested.

Methods

The trial is being carried out in the vicinity of a large urban area. Each model has been established within an area covering a single collection centre. One such centre embraces approximately 200 farms which would usually handle some 1000 cows.

Model 1 – Reference model where no changes to the traditional systems were introduced or stimulated;

Model 2 – Milk cooling equipment distributed to all farmers, improved milk storage facilities at the collection centre, partly insulate transport for cans between farm and collection centre;

Model 3 – Mobile collection centre with bulk tanks on platforms picking up milk from czns placed at farm gates;

Model 4 – Twice-a-day milk collection, cooling at village milk cooling centre, collection by bulk tank lorry to dairy plant;

Model 5 – Large scale farms provided with milk cooling tanks able to hold milk of 2 days' production. Collection by bulk tank every second day.

The models were developed and equipped with financial and technical assistence of the FAO/Poland development project.

Surveys carried out before the implementation of models did not reveal any signifficant differences in milk composition nor quality between the areas (groups of farms) involved. The technologic and economic effectiveness of the various models were tested during summer period of 1985 for the first time. Samples were taken once a week and the physical, chemical and hygienic properties were determined.

Results

The results, which must be regarded as preliminary in the whole experiment, are shown by the table below. In all experimental models an increase of total solids (TS) content has been noted. That increment may be regarded as an improvement of the milk's processing quality. Presumably this could be credited to the influence of intensive advisory activities of the extesion service within the area of experimental models. The higher concentration of butterfat and protein (statistically signifficant), as compared to the reference model, may have resulted from better feeding practices introduced.

It is worthwhile to note a signifficant increase in lactose content in model 2 and fat and protein content in models 3, 4 and 5. In the latter one also an increase in free fatty acids (FFA) content was found, which may indicate at rising lipolytic processes. The comparison of the FFA level in model 5 and model 2 suggests that the lipolysis is rather of bacterial than spontaneous origin. This would require further microbiologic and biochemical investigations.

From the processing technology point of view it is alarming, that in all models some small proportion of milk coagulates in the alizarol test (70% ethanol v/v), hence shows a decreased termostability. This problem, caused most probably by some imbalance in minerals transfer between environment environment and animals will be further investigated. The positive results of bromo-thymol test is a prove of some milk adulteration but also insufficient flashing of detergents and desinfectants from dairy equipment.

Improved milk collection was best reflected in hygienic quality of milk in model 4. More than 70% of milk delivered has been accepted under grade I, 22% under grade II and only 6.67% under the basic grade. The hygienic quality of milk collected under model 3 was even worse than that under reference model.

There is still little evidence yet to pass a judgment on the economic effectiveness of the various models. Further information is being gathered. Preliminary observations indicate that investment into cooling equipment made in model 2 and the arrangements made in model 5 are economically sound under the present price determination system.

The quality of milk in ddifferent models of milk collection
May, June, July, August, September 1985;
No. of samples n = 600
(x = probability 95%; xx = probability 99%)

Characters of raw milk	Model 1	2	3	4	5
Butterfat %	3.74	3.88	3.91x	4.06xx	3.95x
Protein %	3.21	3.32	3.40xx	3.32x	3.36x
Lactose %	4.96	5.07xx	4.93	4.97	4.92
Total solids %	12.61	12.97xx	12.94xx	13.05xx	12.93xx
Pot.acidity SH	7.37	7.24	7.41	7.22	7.30
Act.acidity pH	6.60	6.63	6.62	6.63	6.62
Density g/cm3	1.0312	1.0322xx	1.0319xx	1.0317x	1.0318x
FFA meqv/cm3	0.869	0.944	0.940	0.953	1.173
Somatic cells .000/cm3	651	683	595	666	607
% of samples coagulating in alizarol test	18.6	13.2	9.5	12.8	15.3
Grading hygienic quality - % samples qualified to:					
I gr.(>4 hrs.)	39.29	52.08	22.35	71.11	55.32
II gr. (2-4 hrs.)	34.82	38.19	61.18	22.22	38.46
basic gr.(<2 h.)	25.89	9.73	16.47	6.67	6.22
(reductase test with methylene blue)					

IMPROVEMENT OF MILK PRODUCTION AND QUALITY – FAO PROJECT

M. Jurczak, J. Kwiatkowski, M. Zajac,
Institute of Cattle Breeding and Dairy Science,
Warsaw Agricultural University and Central Union of Dairy Cooperatives, 05-840 Brwinow, Poland

Under the assistance of UNDP and in Cooperation with FAO a small scale project is being carried out in Poland by the Institute mentioned above. Among other objectives (i.e. establishment of training and research facilities) also a research project aiming at the improvement of milk production and its quality is under way. Some results obtained so far are given below.

The specific feature of Polish agriculture is the tremendous dispersion of land among almost 3 million farms. This reflects directly onto the milk production. Over 80 percent of milk delivered originates from private farms, while the remaining 20 percent from large state farms. The cooperative dairy industry purchases milk from approximately 1,5 million farmers, who own a national herd of over 5000000 dairy cows. This makes an average of 3.5 cows per farm. Since the concentration of milk production in private farms is very slow, in the investigation presented an optimal figure of 4 cows/farm has been taken for consideration.

The aim of the research project is to prove to the farmers that application of improved method of milk production and handling may be profitable both, to the dairy industry and to the milk producers.

Material and methods

Within an area near the capital town of Warsaw one thousand private farms with 4 to 5 cows each have been embraced by the experiment. All the farms have been devided into groups within which a different approach to milk handling has been implemented. Here we will show results obtained in one experimental group in comparison to the reference group.

All the farmers of the experimental group have been provided with cooling equipment for chilling the milk on the farm and also with permanent extension service dealing with the improvement of milk production systems.

The initial chemical, physical, higienic and economic surveys did not reveal substantial differences between the groups. During 1983 and 1984 extension work together with the supply of equipment were quite intensive and in 1985 the effectivness of the applied measures was tested.

Results

The figures given in the table present the results obtained. During the 7 months of winter period (October – April) the contnts of lactose , total solids (TS) and density was found to be signifficantly higher in the area where milk was routinely cooled after milking. No difference was found in respect of higienic properties of the milk. Traces of coagulation of protein in alizarol test (70% v/v ethanol), particularly intensive in refence area during the winter period. This reaction was attributed mainly to nonspecific coagulation of milk proteins. During summer period however the difference in coagulation ratio between groups was not that large. Profound differences between the areas compared have been revealed during the 5 summer months (May – September). Not only the effect of extension work but also influence of cooling equipment installed on the farms could be seen. The total effect of 12 percent more milk of top quality delivered (to which a special bonus is added to the farm-gate price) is highly signifficant.

The higienic quality of the milk collected from the experimental area, as determined by the reductase test with methylene blue, has improved considerably. Thus, the extension work must have contributed to the milk technologic quality (increase in TS and density) and hygienic value as well.

Economic studies have proved, that the investment made into cooling equipment and the cost of energy to run them will be returned very quickly (the time depends on the amount of milk produced on the farm) because of the very high difference in prices for top quality raw milk in comparison to the lower grades.

Conclusions

The introduction of permanent and effective extension work at grassroot level together with the provision of farms with the necessery cooling equipment resulted in substantial improvement of technologic and higienic properties of milk from private farms.

Direct benefits went to farmers, who achieved higher prices for their produce due to higher content of milk components and a better hygienic quality grade.

Chemical, physical and hygienic properties of raw milk in two different milk delivery areas and seasons of 1985.
(x = probability 95%; xx = probability 99%)

Characters of raw milk	October-April (n=840)		May-September (n=600)	
	reference	experiment	reference	experiment
Butterfat %	4.30	4.29	3.74	3.88
Protein %	3.39	3.42	3.21	3.32
Lactose %	4.80	5.07xx	4.96	5.07xx
Total solids %	13.19	13.48x	12.61	12.97xx
Potential acidity (gr.SH)	7.04	7.15	7.37	7.24
Active acidity pH	6.65	6.64	6.60	6.63
Density g/cm3	1.0311	1.0323x	1.0312	1.0322xx
Free fatty acids ueqv/cm3	1.351	1.289	0.869	0.944
Somatic cells .000/cm3	627	480	651	683
% of samples with traces of coagula tion in alizarol test	26.9	1.2	18.6	13.2
Grading hygienic quality - % of samples qualified to:				
I grade (>4 hrs.)	75.25	74.77	39.29	52.08
II gr.(2-4 hrs.)	18.81	20.17	34.82	38.19
basic (<2 hrs.)	5.94	7.56	25.89	9.73

(reductase test with methylene blue)

Milk Thermization at the Farm level in Italy

R. Lodi, F. Sangiorgi*, G. Provolo*

Centro latte CNR, *Istituto di Ingegneria Agraria - Università di Milano

Via Celoria, 2 - 20133 Milano, Italy

Foreword

In Italy usually cheese milk is collected twice a day in order to produce Grana and Provolone cheese, while fluid milk is collected every day or every two days, including Saturday and Sunday.

Collection accounts for 2 to 6 % of the cost of milk at the farm. To reduce this cost it is necessary to collect milk at least every 4 days. Since the quality of milk cannot be satisfatorily mantained for more than 2 days only by cooling it at 4°C, a solution to this problem may be a heat treatment of the milk just after the milking plant.

In order to verify the possibility of using this solution under normal conditions a series of farm tests were carried out in Italy, starting in January 1985.

Materials and Methods

The treatment was carried out by using equipment, supplied by Alfa-Laval, based on plate exchangers, that can treat 100 l/h (Fig. 1 and Photo 1). The experiments were carried out at the experimental farm of the University of Milan.

In order to choose the optimum temperature, a series of tests were first carried out out on raw milk heated and kept for 15 s at different temperature levels (from 64°C up to 80°C) before cooling it at 4°C.

Samples for analyses were taken from raw milk and heated milk.

The following Bacteriological and Rheological analyses were carried out:

- Standard plate count were made in Plate Count Agar (Difco). Incubation: 32°C for 72 h;
- Psychrotrofs were counted in the same medium after 10 days of incubation at 7°C;

- Coliforms were counted in Violet Red Bile Agar (Difco). Plate were incubated at 37°C for 24 h;
- Coagulation properties of milk were measured with lactodynamograph.

Afterwards the following experiments were carried out for the treatment with the fixed parameters of 70°C for 15 s:

- Control of bacteriological and rheological characteristics of the milk during 7 days of refrigerated storage after the treatment (without adding new milk);
- Control of bacteriological and rheological characteristic of the milk during 7 days of refrigerated storage adding new treated milk at each of the two daily milkings for the seven days.

  This last test was carried out both in good hygienic conditions and in less well controlled conditions.

Results

The results concerning the choice of the treatment temperature are presented in fig. 2.

The results of this second series of tests are summarized in fig. 3, 4, 5.

Final considerations

From the results of these experiments it may be pointed out that:

1 - The heat treatment efficiently reduces SPC and Psycrotrophs, and eliminates Coliforms;

2 - The coagulation properties of the milk are reduced to an acceptable extent but not so drastically to make trouble for dairying.

3 - Thermization cannot replace good hygiene in the handling of milk on the farm, before and especially after the heat treatment. In fact new contamination is possible due either to the lower microbiological competition or to the long period of storage.

    This fact is more evident if it is considered that the possible users of this equipment would be the small or middle sized farms;

4 - The energy required for the treatment was 20 Wh/l.

Research carried out in the frame of PFE of the National Reasearch Council. Contract n. 83.02929

EFFECTS OF GENETIC FACTORS ON MILK YIELD UNDER THE CONDITIONS OF LARGE-SCALE PRODUCTION

DOC.ING.J.PYTLOUN,CSc., ING.J.MOTYČKA,CSc.,/Agricultural University Prague/,ING.F.URBAN,CSc.,ING.J.BOUŠKA,CSc./ResearchInstitute of Animal Production Prague -Uhříněves/,ING.J.VETÝŠKA,CSc./State Breeding Enterprises Hradištko/.

609 first-calf heifers with average milk yield of 4052 kg kept in litter houses with stanchions up to 200 heads each were studied. Identical rations were fed. They were prepared in a central mixing plant for roughage;concentrates were fed according to performance.
Milk yield in 9 different genotypes obtained from grading up the native Czech Spotted breed with Black and White bulls. The data obtained from the first milking correspond more with the results from a standard lactation than with those from a shortened 100-day lactation. Holstein-Friesian bulls showed a positive effect on the level of milk yield. All the differences were statistically insignificant, even though in the standard lactation they amounted up to 16.43 %.
Primipara milk yield in dependance on the level at which their fathers were tested by progeny testing.The results show that for the set given /milk yield of approximately 4000 kg/the sires should be used when proved at the level of 3500 kg of milk and more. The milk yield of 4360 kg achieved in young tested bulls may be evaluated as positive /group 5/. In contrast to it,the milk yield of daughters of sires proved at the level to 3000 kg,resp.3500 kg is about 600 kg lower than in the daughters of sires proved at the level of 3501-4000 kg,resp.4001 kg and more/first-calf heifers milk yield of about 4500 kg/ for a standard lactation.Again the results from the first milking approach more the evaluation of the standard lactation than that of the shortened one. The results obtained are statistically significant.
First-calf heifer milk yields in dependance on relative breeding values of sires for milk production established on the basis of progeny testing results was studied. First-calf heifer milk yields on the basis of CC-test /for milk/ of the sires tested by progeny testing.The values of first-calf heifer milk yield in dependance on the origin after father/a total of 10 fathers was evaluated/.

EFFICACY OF A 'COLD CLEANING' SYSTEM FOR FARM MILKING PLANTS

J BRUCE
MICROBIOLOGY DEPARTMENT
WEST OF SCOTLAND AGRICULTURAL COLLEGE
AUCHINCRUIVE
NEAR AYR
SCOTLAND KA6 5HW

INTRODUCTION

The cost of heating wash water is a significant element in the total costs of cleaning farm milking plants. The development of a cleaning system which is less dependent on hot water is therefore of interest. Such a system has recently been introduced for use on farms in the U K and is the subject of the trial described below in which the 'cold cleaning' system is compared with a conventional hot wash cleaning system over a nine month period on a 16 x 16 herringbone parlour.

DETAILS OF TRIAL

STANDARD HOT WASH

Applied after morning and evening milkings. Pre-rinse to remove milk residues. Wash - hot water, plus proprietory detergent-disinfectant. Final rinse - cold water, plus hypochlorite.

COLD CLEANING SYSTEM

The active component in the cold cleaner is a dihalogen - iodine monochloride stabilised in nitric acid solution.

The cold cleaning system consisted of six post-milking washes using the cold cleaner, followed by a conventional hot wash after every seventh milking. Thus, in the course of one week the plant received twelve cold cleans and two hot cleans.

The hot wash element of the cold cleaning system employed in the major part of the trial was a proprietory brand of detergent-disinfectant powder. In the latter part of the trial a liquid detergent-disinfectant was used as the hot wash element in the cold cleaning system (see Table 1).

a) Cold wash element

Pre-rinse to remove milk residues. Wash - cold water, plus cold cleaner. Final rinse - cold water, plus hypochlorite.

b) Hot wash element

As described above for standard hot wash.

The two wash procedures were alternated during the trial as indicated in Table (1).

Table 1 SEQUENCE OF WASH PROCEDURES DURING TRIAL

Standard Hot Wash	Jan	-	Feb	24 days
Cold Cleaning System (P)	Feb	-	March	19 days
Standard Hot Wash	March	-	May	71 days
Cold Cleaning System (P)	May	-	Aug	71 days
Cold Cleaning System (L)	Aug	-	Sept	36 days
Standard Hot Wash	Sept	-	Oct	32 days
				253 days

(P) = Powdered detergent disinfectant used in hot wash element of the cold cleaning system.

(L) = Liquid detergent disinfectant used in hot wash element of the cold cleaning system.

TEST PROCEDURES

The bacteriological quality of the plant was assessed using the rinse technique described in BS 5226. Rinse and milk samples were taken thrice weekly on Mondays, Tuesdays and Fridays as indicated in Table (1). The results are expressed as $\log_{10}$ of the number of bacterial per square metre of plant. Milk samples were examined for total bacterial count at 30° as described in BS 4285.

RESULTS

The visual cleanliness of the plant was maintained satisfactorily throughout the trial. A slight deposit thought to be of protein origin was formed on the jars and ACR units during the first usage of the cold cleaner. However, this depost was removed by the hot wash element of the cold cleaning system and was not observed subsequently.

The results presented in Fig (1) and summarised in Table (2) indicate that washing with the Cold Cleaning System did not adversely affect the bacteriological quality of the plant. Furthermore, the rinses obtained when the standard hot wash was in use tended to be higher than those obtained during use of the Cold Cleaning System.

The level of bacterial contamination in the milk was not affected in any observable way by the cleaning sytems studied.

PLANT RINSE RESULTS FIG (1)

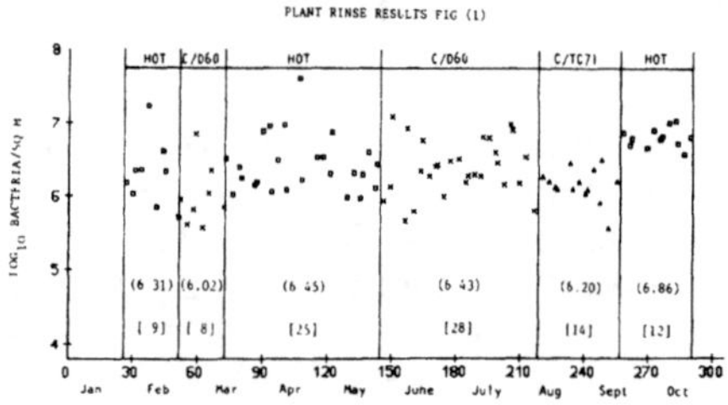

HOT = STANDARD HOT WASH. COLD/P = COLD CLEANING USING POWDERED DETERGENT/DISINFECTANT IN THE HOT ELEMENT. COLD/L = COLD CLEANING USING LIQUID DETERGENT/DISINFECTANT IN THE HOT ELEMENT. ( ) = MEAN LOG_{10} BACTERIA/SQm. [ ] = NUMBER OF SAMPLES

Table 2 SUMMARY OF PLANT RINSE RESULTS

Type of Wash	No of Rinses	Mean $\log_{10}$/Sq m
Standard Hot	46	6 53
Cold Cleaning System (p)	36	6 340
Cold Cleaning System (L)	14	6 200

Department of Microbiology, The West of Scotland Agricultural College, Auchincruive, Ayr, KA6 5HW

SIMULATOR EVALUATION OF HYGIENE IN NEW ZEALAND

D.G. DUNSMORE, D.M. BLACKLOCK AND G.H. DERRICK
NATIONAL DAIRY LABORATORY, RUAKURA ARC, PRIVATE BAG, HAMILTON, NEW ZEALAND.

Simulators are automated machines which repeatedly apply a soiling solution (milk in our studies) and the components of the cleaning system to test surfaces. Over many soil/wash cycles the soil accumulates and bacteria grow on the surfaces. They can be used to examine how cleaning systems operate, develop commercial systems or to examine the hygienic performance of milk-contact equipment (1).

There are currently 3 simulators at the NDL. A Rubberware Simulator (1) has been used for some years to examine the performance of pulsating teat cup liners. A new Milking Machine Component Simulator has recently been constructed to examine the hygienic performance of such components as clusters, filters, meters and coolers. A Hygiene Simulator has also recently been completed which has replaced the pipeline machine previously used (2).

There are also three other simulators which were in use in Australia. They are being relocated to the NDL. One of those machines is ued to examine the performance of cleaning systems for vats (3). The second machine is a soak machine (4), modified to also examine flow systems. The third machine is a multipot dipping machine (5), which will be used to examine new construction materials.

References

(1) Salvatierra, S.A.: 1973. Proc. NZ Sch. Conf. Milk Quality 42. University of Auckland.

(2) Murphy, B.T.: 1973. Proc. NZ Sch Conf. Milk Quality 57. University of Auckland.

(3) Dunsmore, D.G. and Barnes, R.N.; 1978. Aust. J. Dairy Technol. 33:114.

(4) Dunsmore, D.G.; Westwood, D; Jay, D. and Embling, M.; 1980. J. Food Prot. 43:15.

(5) Dunsmore, D.G.; Barnes, R.N.; Scott, J.M. and Wheeler, R.G.; 1978. Aust. J. Dairy Technol. 33:100.

A SURVEY OF MILKING MACHINE CLEANING PRACTICES IN NEW ZEALAND

D.G. DUNSMORE AND P. TOMLINSON,
NATIONAL DAIRY LABORATORY, RUAKURA ARC, PRIVATE BAG,
HAMILTON, NEW ZEALAND.

The aim of this study was to document the machine designs and cleaning practices and relate those to the bacteriological status of the machine surfaces and the milk produced on the farm, by surveying 205 New Zealand pipeline milking machines. The machine designs and cleaning practices were obtained by questionnaire, and the hygienic condition by bacterial swabs on machine surfaces, observation (FDI rating) and by quality evaluation of milk received of the factory. One-way analyses of variance were conducted on the relationships between the machine design or cleaning system; and each of the hygiene parameters. Some of the inforamtion has been previously reported (1).

The performance of two cleaning systems recommended by government advisors was very good. However, a number of simplifications of those systems were also found to perform satisfactorily. The hygienic status of machines cleaned by systems recommended by governmental farm advisors was superior to those cleaned by systems recommended from other sources.

Machines cleaned by reverse-flow were less hygienic than both recirculation and flush-cleaned machines. The hygienic status of the machines worsened with increasing machine age and also decreasing replacement frequency of teat cup liners. The hygienic status of machines cleaned by systems recommended by governmental farm advisors was superior to those cleaned by systems recommended from other sources.

The swab grades of bacteria on the machine surfaces, the methylene blue test and the FDI's hygiene rating were all significantly correlated. The thermoduric count and freezing point test were not significantly corrected to any other hygiene parameter.

Reference (1) Tomlinson, P.; Dunsmore, D.G. and Cox N.R; 1981. NZ J. Dairy Sci. Technol. 16:273.

HIGHER MILK YIELDS AND IMPROVED UDDER HEALTH THROUGH THE USE OF AUTOMATIC MILKING CLUSTER REMOVERS WITH STRIPPING EFFECT

D. Bothur, W. Ebendorff[+], H.-J. Rudovsky, M. Graupner
Kombinat Fortschritt Landmaschinen, VEB Anlagenbau Impulsa, 7904 Elsterwerda/[+]Institute of Cattle Production, Academy of Agricultural Sciences, 3541 Iden/ German Democratic Republic

The influence of stripping on milk yield and udder health was analysed. Half udder tests showed increasing milk and fat losses from 1st up to 4th lactation and worse udder health in the non-stripped udder halfes. Automatic removal of the teatcups without stripping causes problems. A new automatic cluster remover with stripping effect proves as an alternative.

MONDAY - POSTER 9

DEVELOPMENT OF DAIRY INDUSTRY IN CHINA

Guanglie Han, Liquan Huang[*]
Food Industry Bureau, Ministry of Light Industry, Beijing, China.
[*]Heilongjiang Dairy Industry Research Institute, Harbin, P.R. China.

Foundation of the Dairy Industry in China was relatively poor, but now it shows a high rate of development. There are 1.1 million dairy cows in China. Some goats and local cows are also used for milking. The principle dairy products are milk powder, baby food and pasteurized milk.

RHEOLOGISCHE EIGENSCHAFTEN DER BUTTER JE NACH DER FETTZUSAMMENSETZUNG UND DEN HERSTELLUNGSVERFAHREN

J. Kisza, B. Staniewski, K. Batura, K. Borewski
Institut für Milchwirtschaftliche Technologie
der Landwirtschafts-Technischen Akademie
Olsztyn, Polan

Je nach die Jahreszeit wurden die Unterschiede in chemischer Fettzusammensetzung und die folgende daraus Unterschiede in der Jodzahl sowie auch in der dilatosotrischen Charakteristik festgestellt. Die Untersuchungen zeigen eine wesentlichen Einfluss der Fettzusammensetzung auf die rheologische Eigenschaften der Butter von der kontinuierlichen und periodischen Verfahren.

THE INFLUENCE OF THE WHEY PROTEIN DENATURATION ON THE FLOW PROPERTIES OF MILK ULTRAFILTRATION CONCENTRATES

DIPL.-ING. ITAMAR C. DE CARVALHO and o.PROF. DR.-ING. H.G. KESSLER

Institute for Dairy Science and Food Process Engineering,
Technical University of Munich,
D-8050 Freising-Weihenstephan

INTRODUCTION

THE PRODUCTION OF UF-CHEESE IS ALREADY IN SOME COUNTRIES WELL ESTABLISHED. HOWEVER, THE KNOWLEDGE OF IMPORTANT ENGINEERING PARAMETERS (AND THEIR INFLUENCING FACTORS) SUCH AS THE VISCOSITY OF THE CONCENTRATES IS STILL LACKING. THIS INFORMATION IS OF PARAMOUNT IMPORTANCE FOR THE CORRECT DESIGN OF PROCESSING EQUIPMENT AND PIPE-LINES.

THE FLOW BEHAVIOUR OF THESE PRODUCTS IS NON-NEWTONIAN (PSEUDOPLASTIC), I.E. THE RELATION BETWEEN SHEARING STRESS τ AND VELOCITY GRADIENT (dw/dy) IS NOT CONSTANT.

THE POWER LAW FROM OSTWALD $\tau = K_{ow}(dw/dy)^n$ CAN DESCRIBE THE FLOW PROPERTIES OF TIME-INDEPENDENT NON-NEWTONIAN LIQUIDS. IT IS WIDELY USED IN ENGINEERING CALCULATIONS. K_{OW} = CONSISTENCY INDEX IN $Pa.s^n$ n = FLOW BEHAVIOUR INDEX

THE FLOW BEHAVIOUR INDEX IS READILY DETERMINED AS THE SLOPE OF THE PLOT OF τ VERSUS (dw/dy) ON LOGARITHMIC COORDINATES. THE VALUE OF n IS LESS THAN UNIT FOR PSEUDOPLASTICS. IF $n = 1.0$ THAN $K_{OW} = \eta$ I.E. THE VISCOSITY.

THE VELOCITY PROFILE OF A FLUID WILL VARY ACCORDING TO VARIATIONS OF THE FLOW BEHAVIOUR INDEX n. PRESSURE DROP DEPENDS BOTH ON THE CONSISTENCY INDEX K_{OW} AND ON THE VALUE OF THE FLOW BEHAVIOUR INDEX.

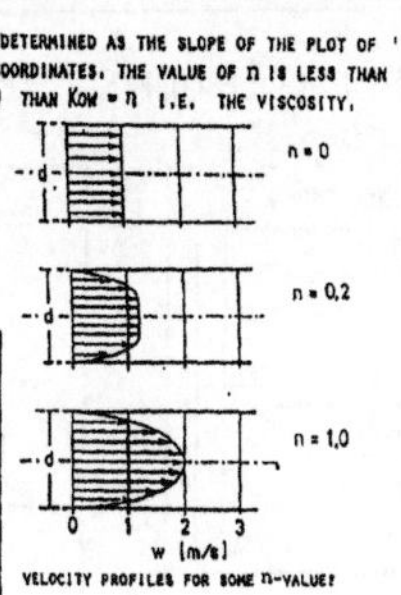

VELOCITY PROFILES FOR SOME n-VALUES

OBJECTIVES

BEARING THESE PRINCIPLES IN MIND IT WAS TRIED TO QUANTIFY THE MODIFICATIONS CAUSED BY THE HEAT TREATMENT OF THE MILK I.E. BY THE WHEY PROTEIN DENATURATION ON THE FLOW PROPERTIES OF THE PRODUCED CONCENTRATES.

MATERIALS AND METHODS

THE PRODUCT WAS SKIMMILK. THE FLOW CURVES WERE DETERMINED WITH A ROTATIONAL VISCOMETER AT 20 C. DRY MATTER CONTENT VARIED BETWEEN 14.0 AND 30.0%. SHEAR GRADIENT WAS VARIED BETWEEN 5 TO 3687 s^{-1}. THE MILK WAS HEATED AT DIFFERENT TEMPERATURE/TIME COMBINATIONS IN ORDER TO ACHIEVE PROTEIN DENATURATION RATES IN TERMS OF β-LACTOGLOBULIN B IN AN RANGE OR 1 TO 99%.

THE EFFECT OF WHEY PROTEIN DENATURATION WAS STUDIED BY MEANS OF THE INCREASE OF THE CONSISTENCY INDEX K_{OW}.

ACCORDING TO THE KINETICS OF HEAT DENATURATION OF PROTEINS, BUT BEARING IN MIND THAT THE VISCOSITY OF THE CONCENTRATES INCREASES AS A RESULT OF DENATURATION, THE FOLLOWING TIME LAW WAS DERIVED.

$$\frac{K_{ow(t)}}{K_{ow(o)}} = \left[1+(r+1).k_f.t\right]^{\frac{1}{r+1}}$$

r = reaction order (-)
k_f = temperature dependent velocity constant (1/s)

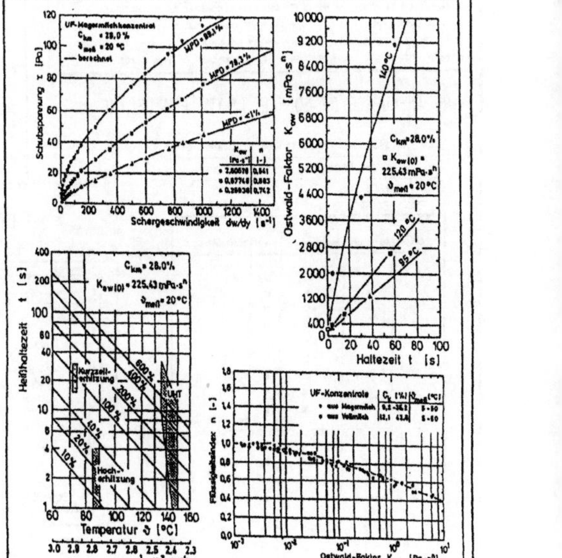

SUMMARY

THE INFLUENCE OF WHEY PROTEIN DENATURATION (RESULTING FROM THE HEAT TREATMENT OF THE MILK) ON THE FLOW PROPERTIES OF UF-CONCENTRATES WAS STUDIED. WHEREAS THE EFFECT ON MILK VISCOSITY IS NEGLIGIBLE, THE FLOW PROPERTIES OF THE PRODUCED CONCENTRATES ARE STRONGLY AFFECTED. AN IMPORTANT ASPECT OF THESE MODIFICATIONS IS THAT THEY ARE MORE INTENSE WHEN THE DRY MATTER CONTENT IS INCREASED.

THE MODIFICATIONS CAN BE QUANTITATIVELY DESCRIBED UNDER THE LIGHT OF REACTION KINETICS.

THE RESULTS MAKE IT POSSIBLE TO OPTIMIZE HEATING PROCESSES BEFORE ULTRAFILTRATION. THE FINAL FLOW PROPERTIES CAN BE PREDICTED IN ORDER TO AVOID EXTREMELY HIGH VISCOSITIES OF THE CONCENTRATES. IN CASE THESE HIGH VISCOSITY VALUES ARE DESIRED, THE FLOW BEHAVIOUR CAN BE PREVIOUSLY DETERMINED AND POSSIBLE MODIFICATIONS ON THE PROCESSING EQUIPMENT CAN BE UNDERTAKEN.

INSTITUTE FOR DAIRY SCIENCE AND FOOD PROCESS ENGINEERING
TECHNICAL UNIVERSITY OF MUNICH D-8050 FREISING/WEIHENSTEPHAN

PROPERTIES OF COFFEE AND COFFEE CREAM WITH RESPECT TO STABILITY OF COFFEE CREAM

Dipl.-Ing. Sabine Geyer and Prof.Dr. H.G. Kessler
Institute for Dairy Science and Food Process Engineering, T.U.-Munich
D-8050 Freising-Weihenstephan

INTRODUCTION

BOTH THE COFFEE- AND THE MILK INDUSTRIE ARE FAMILIAR WITH THE PROBLEM OF THE FEATHERING OF MILK/CREAM IN HOT COFFEE. TILL NOW HOWEVER THERE ARE NO UNEQUIVOCAL RESULTS, TO WHAT EXTENT THE COFFEE AS WELL AS THE COFFEE CREAM ARE RESPONSIBLE FOR THE FEATHERING.
THE OBJECTIVE OF THIS PROJECT IS ACCORDINGLY, ON THE ONE SIDE TO DETERMINE THE INFLUENCES OF THE COFFEE BEVERAGE ON THE FEATHERING AND ON THE OTHER SIDE TO CHANGE THE TECHNOLOGY OF THE PRODUCTION OF CREAM IN ORDER TO AVOID THE FEATHERING OF CREAM IN HOT COFFEE.

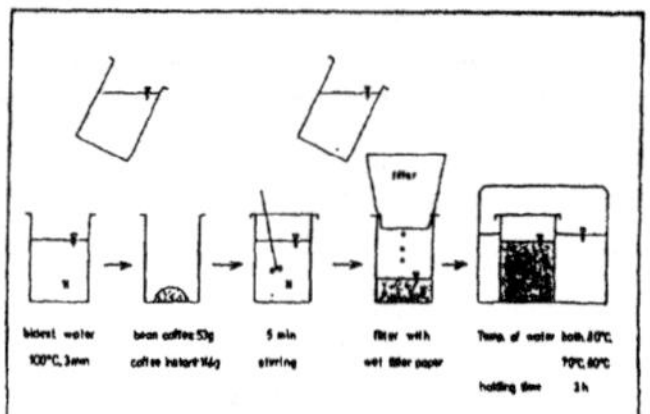

Standard Method of Coffee Preparation

CREAM STABILITY TEST:
ADDING 0.3 ML COFFEE CREAM TO 10 ML HOT COFFEE (80°C, 70°C, 60°C) AND ESTIMATING VISUALLY THE DEGREE OF FEATHERING:

DEGREE OF FEATHERING 1 = NO FEATHERING
DEGREE OF FEATHERING 2 = NOT VERY INTENSE
DEGREE OF FEATHERING 3 = INTENSE
DEGREE OF FEATHERING 4 = VERY INTENSE
DEGREE OF FEATHERING 5 = EXTREMELY INTENSE

THE INFLUENCE OF COFFEE ON THE FEATHERING

COMPARISION OF DIFFERENT COFFEE BRANDS BY MEANS OF THE CREAM STABILITY TEST (WITH JUST ONE CREAM)

CATEGORY	COFFEE BRAND	pH-VALUE $_{80°C, 3h}$	ACID NO.	DEGREE OF FEATH.
I	A	4.62	80	5
	B	4.66	83	5
	C	4.67	87	5
	D	4.68	86	5
	E	4.70	80	5
	F	4.73	78	3
	G	4.75	85	4
	H	4.76	80	4
	I	4.80	73	3
II	J	4.63	62	3
	K	4.70	60	3
	L	4.68	60	3
III	M	4.91	66	2
	N	4.89	59	1.5
	O	4.90	65	2

THE INFLUENCE OF THE TOTAL ACID CONTENT ON THE FEATHERING OF CREAM:

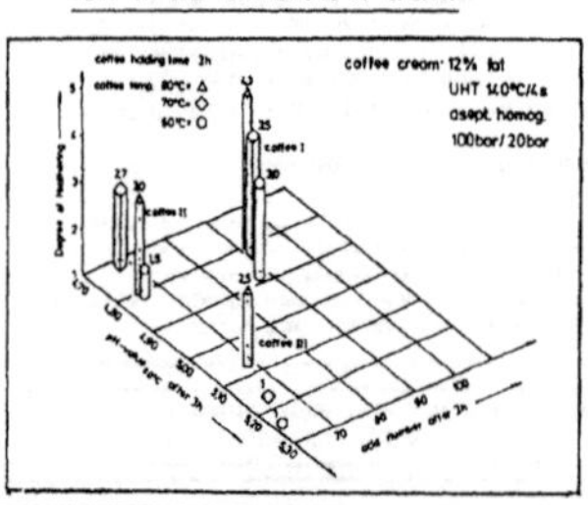

THE INFLUENCE OF COFFEE CREAM ON THE FEATHERING

Influence of the Homogenization on the Feathering

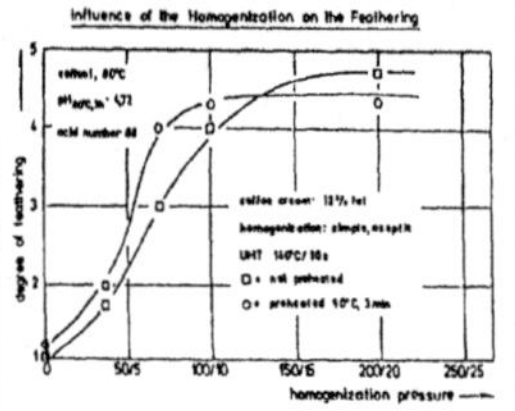

Influence of the Heating on the Feathering

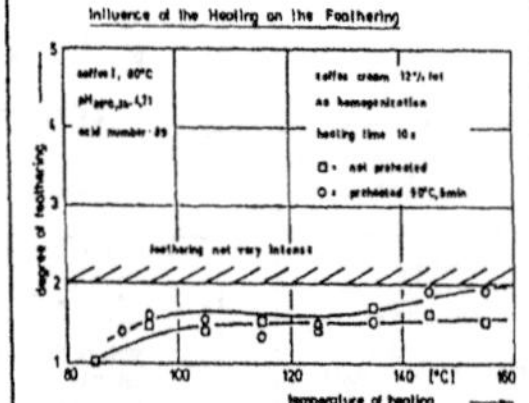

Influence of the Homogenization Process on the Feathering

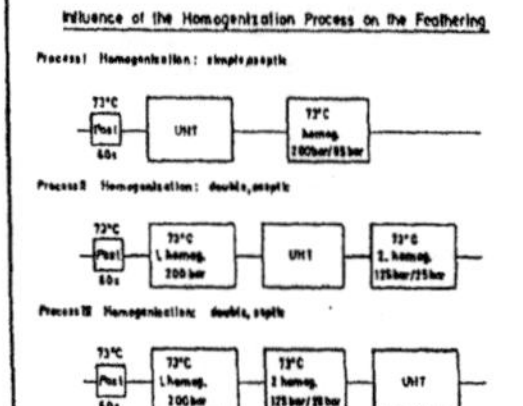

Influence of the Homog. Process and the Heating on the Feathering

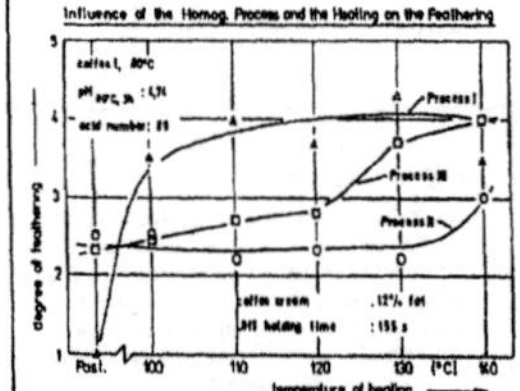

SUMMARY

COFFEE BEVERAGE AS WELL AS COFFEE CREAM ARE RESPONSIBLE FOR THE FEATHERING. DEPENDENT ON ITS PROCESSING TECHNOLOGY THE COFFEE MAY POSSESS A HIGH TOTAL ACID CONTENT WHICH AFFECTS THE STABILITY OF CREAM IN HOT COFFEE. HOMOGENIZATION AND HEAT TREATMENT CONDITIONS DURING THE PRODUCTION PROCESS OF COFFEE CREAM PLAY AN IMPORTANT ROLE ON THE COLLOIDAL PROTEIN STABILITY AND THEREFORE ON THE FEATHERING.

INSTITUTE FOR DAIRY SCIENCE AND FOOD PROCESS ENGINEERING, D-8050 FREISING-WEIHENST.

OPTIMUM HOMOGENIZING CONDITIONS FOR CREAMS WITH DIFFERENT FAT CONTENT.

DIPL.-ING. H. EIBEL and o.PROF. DR.-ING. H.G. KESSLER

Institute for Dairy Science and Food Process Engineering,
Technical University of Munich,
D-8050 Freising-Weihenstephan

INTRODUCTION

THE OBJECTIVE WAS TO IMPROVE THE EMULSION STABILITY OF 15 TO 45% FAT CREAM (WITHOUT ADDITIVES) BY MEANS OF OPTIMAL HOMOGENIZING CONDITIONS, WHICH ARE EXPECTED TO BE VERY DIFFERENT BECAUSE OF THE WIDE RANGE OF FAT CONTENTS; ADDITIONALLY TO SHOW THE EFFECT OF THESE PARAMETERS ON THE CONSISTENCY OF CREAM USED FOR DIFFERENT PURPOSES AND TO SHOW THE DEPENDENCE BETWEEN STABILITY AND CONSISTENCY.

MATERIALS AND METHODS

CREAM: PASTEURIZED CREAM (IN CASE OF 15% ALSO UHT-TREATED) WITH FAT CONTENTS OF 15, 25, 30, 36 AND 45% WAS USED.

HOMOGENIZER: A PILOT PLANT HAVING A KNIFE-EDGE LIKE VALVE DESIGN IN BOTH STAGES WITH A CAPACITY OF 80L/H WAS USED.

STEPS OF OPTIMIZATION

* PRESSURE P_1 BEFORE THE 1ST STAGE (ONE-STAGE HOMOG., TEMP. 70°C)
* PRESSURE RATIO P_2/P_1 (TEMP. 70°C): FUNCTION OF BACK PRESSURE AND 2ND STAGE
* EFFECT OF TWO-STAGE HOMOG. (OPTIMAL P_2/P_1; 70°C) IN CONTRAST TO ONE-STAGE HOMOG. OVER A WIDE RANGE OF PRESSURES P_1
* TEMPERATURE AT VARIOUS PRESSURES P_1 (ONE- AND TWO-STAGE HOMOG.)

ANALYSIS

* MEAN DROPLET DIAMETER $D_{50,3}$ AS A CHARACTERISTIC OF THE VOLUME SIZE DISTRIBUTION
* HOMOGENIZING EFFECT Z_A AS A CREAMING PARAMETER IN A CENTRIFUGAL FIELD (300G, 210 MIN, 20°C)

$$Z_A = \frac{\text{MEAN FAT CONT. OF THE LOWER 20\% OF THE TUBE}}{\text{MEAN FAT CONTENT OF THE WHOLE SAMPLE}} \cdot 100\%$$

* APPARENT VISCOSITY (20°C) AT LOW SHEAR RATES (MOSTLY $D = 4.52\ s^{-1}$)

RESULTS:

APPROXIMATELY A PRESSURE RATIO P_2/P_1 OF 0.20 IS IN GENERAL OPTIMUM WITH THE POSSIBLE VARIATIONS BETWEEN 0.15 AND 0.25.

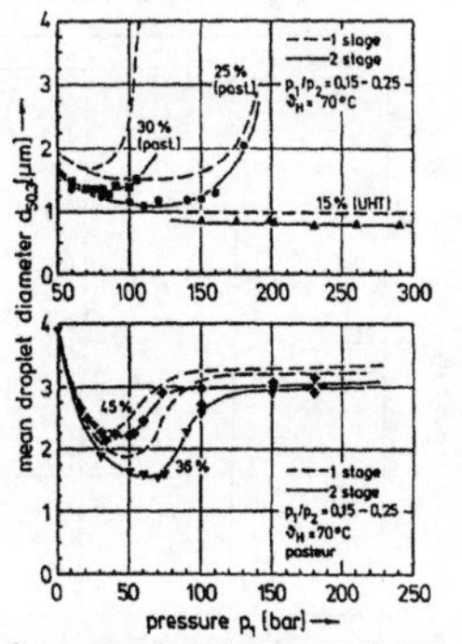

FIG.1: INFLUENCES ON THE MEAN DROPLET DIAM.

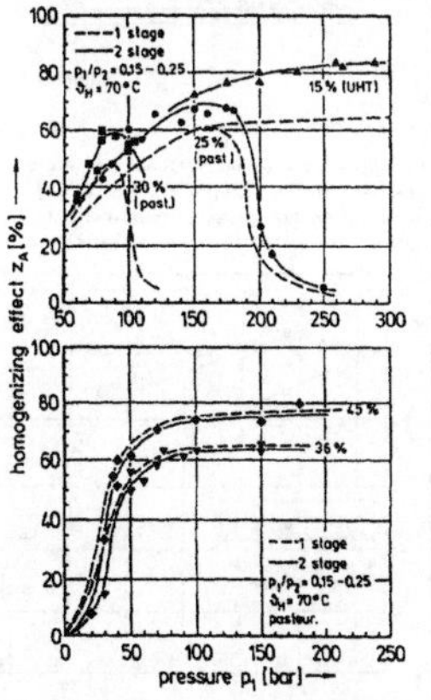

FIG.2: INFLUENCES ON THE HOMOG.-EFFECT

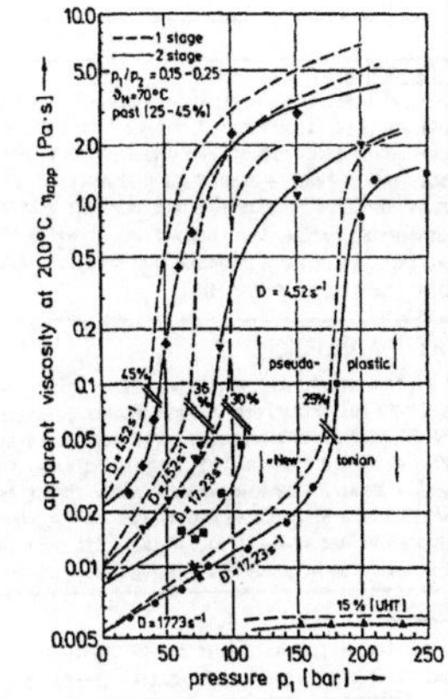

FIG.3: INFLUENCES ON THE APP. VISCOSITY

CONCLUSIONS

* THE MOST EFFECTIVE TEMPERATURE FOR MAXIMUM STABILITY AND MINIMUM VISCOSITY RANGES BETWEEN 75°C (15% FAT) AND 65°C (45% FAT). HOWEVER, IN GENERAL 70°C MAY BE SELECTED WITHOUT REMARKABLE DECREASE IN STABILITY AND INCREASE IN VISCOSITY.
* TO ACHIEVE THE SMALLEST MEAN DROPLET DIAMETER AS THE FAT CONTENT INCREASES FROM 15% TO 45% THE PRESSURE HAS TO BE LOWERED FROM ABOUT 300 BAR TO 40 BAR.

  HOWEVER TO ACHIEVE MAXIMUM STABILITY (= HOMOGENIZING EFFECT) IT IS RECOMMENDED THAT THE PRESSURE SHOULD BE LOWERED ONLY UP TO 90 BAR (MOST SUITABLE FOR 30% FAT). AS THE FAT CONTENT INCREASES ABOVE 36% THE PRESSURE SHOULD NOT BE LOWER THAN 100 BAR. UNDER THESE CONDITIONS, HOWEVER, THE VISCOSITY INCREASES TO A HIGH EXTENT.
* BY HOMOGENIZING 15 TO 30% FAT CREAM WITH TWO STAGES (APPLYING THE MOST EFFECTIVE PRESSURE RATIO OF 0.20) BOTH REMARKABLE INCREASE IN STABILITY AND DECREASE IN VISCOSITY CAN BE OBSERVED. AS THE FAT CONTENT INCREASES ABOVE 36% IT RESULTS ONLY IN DECREASE IN VISCOSITY.

  THESE EFFECTS ARE NOT ONLY DUE TO THE SUITABLE BACK PRESSURE BUT ALSO DUE TO THE DISRUPTION OF CLUSTERS CAUSED BY THE SECOND STAGE.

WHEY PROTEIN DENATURATION AND DEPOSIT FORMATION AT HOT SURFACES.

F. DANNENBERG AND H.G. KESSLER

Institute for Dairy Science and Food Process Engineering,
Technical University of Munich,
D-8050 Freising Weihenstephan.

INTRODUCTION

DEPOSIT FORMATION AT HOT SURFACES DURING HEATING OF WHEY AND MILK IS KNOWN TO BE HIGHLY INFLUENCED BY AGGREGATION REACTIONS OF WHEY PROTEINS. PREHEATING RESULTING IN PROTEIN DENATURATION REDUCES THE AMOUNT OF DEPOSIT. OBJECTIVE WAS TO STUDY RELATIONS BETWEEN THE FOULING REACTION AND DENATURATION OF β-LACTOGLOBULIN WHOSE DENATURATION KINETICS HAVE BEEN INVESTIGATED PREVIOUSLY.

MATERIALS AND METHODS

A PILOT HEATING PLANT WITH SMALL EXPERIMENTAL TUBES WAS USED FOR INVESTIGATING FORMATION OF DEPOSITS DURING HEATING OF SKIM MILK. DIFFERENT PARAMETERS SUCH AS HEATING TEMPERATURE, PREHEATING TIME, PH-LEVEL, RUNNING TIME WERE VARIED. AMOUNT OF DEPOSIT WAS MEASURED BY WEIGHING AND RELATED TO THE CONCENTRATION OF NATIVE β-LACTOGLOBULIN IN THE BULK. AN EVALUATION OF KINETICS WAS CARRIED OUT.

RESULTS

DEPOSIT FORMATION AS A FUNCTION OF DENATURATION RATE OF β-LACTOGLOBULIN AND HEATING TEMPERATURE.

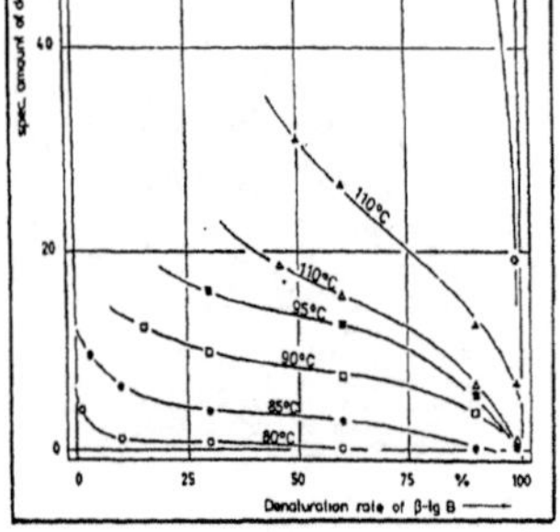

DEPOSIT FORMATION AS A FUNCTION OF RUNNING TIME AND DENATURATION RATE OF β-LACTOGLOBULIN.

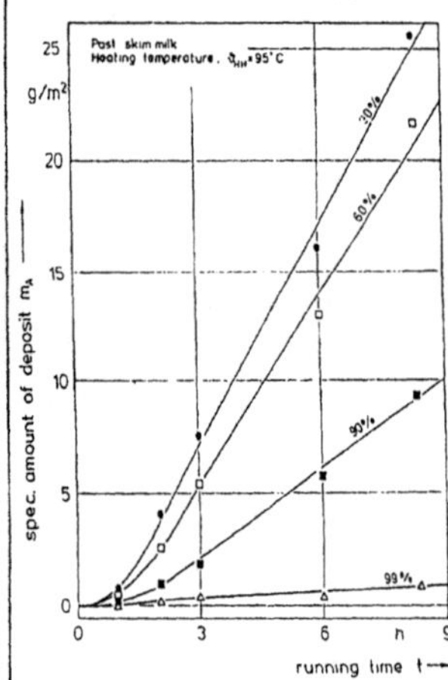

RATE CONSTANT K OF IRREVERSIBLE WHEY PROTEIN DENATURATION PLOTTED AS A FUNKTION OF 1/T, ACCORDING TO THE ARRHENIUS EQUATION (DATA DERIVED FROM PREVIOUS REACTION KINETIC STUDIES).

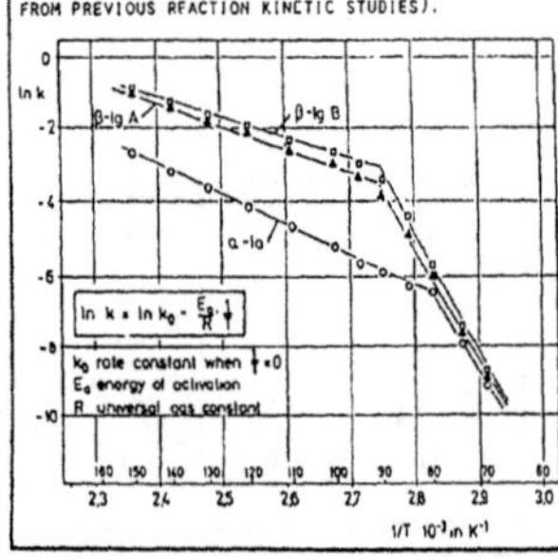

RATE CONSTANT K_p OF THE PROTEIN FOULING REACTION AT DIFFERENT DENATURATION RATES OF β-LACTOGLOBULIN AS A FUNCTION OF 1/T.

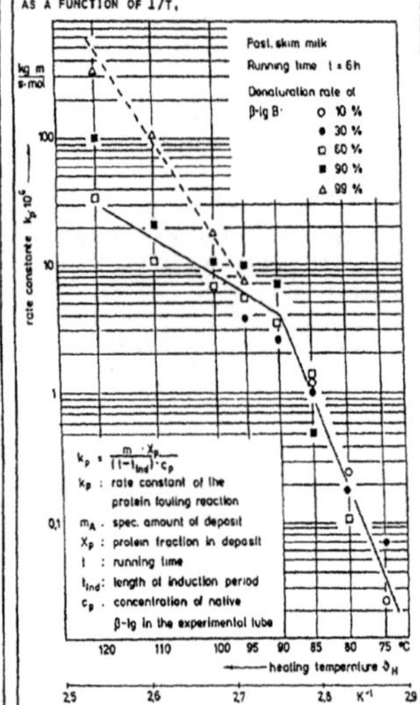

SUMMARY

DEPOSIT FORMATION DURING HEATING OF SKIM MILK CAN BE REDUCED CONSIDERABLY BY PREHEATING OR ADJUSTING PH-LEVEL IN THE BULK (NOT SHOWN). THE MECHANISM OF FOULING BETWEEN 75 AND 110°C SEEMS TO BE RELATED TO THE DENATURATION OF β-LACTOGLOBULIN, SINCE REACTION KINETICS ARE VERY SIMILAR TO THOSE OF IRREVERSIBLE DENATURATION OF β-LG. THUS, FOR LIMITING DEPOSIT FORMATION, CONCENTRATIONS OF NATIVE WHEY PROTEINS AND PROTEIN AGGREGATION SHOULD BE REDUCED.

INSTITUTE FOR DAIRY SCIENCE
AND FOOD PROCESS ENGINEERING
TECHNICAL UNIVERSITY OF MUNICH
D-8050 FREISING-WEIHENSTEPHAN

A CENTRIFUGE WITH A NEW METHOD OF SEPARATED SOLIDS DISCHARGE

H. Komsta, V.D. Surkov +
Technical University, Lublin, Poland
+Moscow's Technology Institut of Meat and Milk Industry, Moscow, USSR

In order to simplify separated solids discharge from separators of efficiency to 5 m^3/h an experimental separator has been constructed which works along the following pattern: bowl corresponds to the rotor of an electric engine, the disos stack with blades attached on the side surface, is placed with its conical part towards the bottom and it is driven irrespectively of the bowl frame.

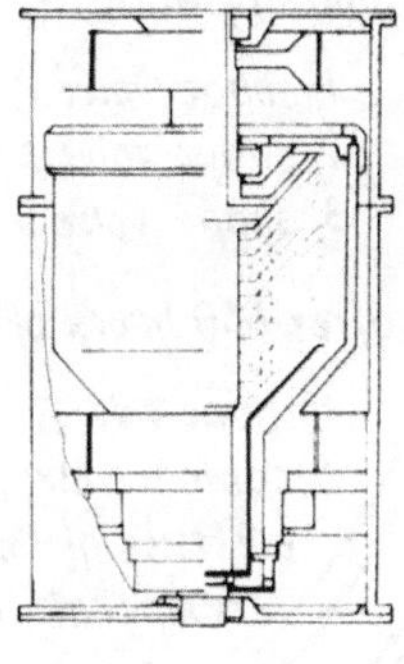

The section of an experimental separator

The general sight of an experimental separator

The elementary separator units: 1 bowl frame, 2 discs stack 3 upper frame with the discs stack stator, 4 bottom frame with bowl frame stator.

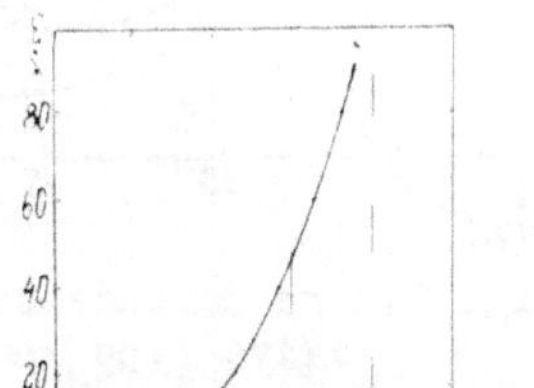

The influence of rotation difference between the discs stack and the bowl frame Δn on the value of the convection coefficient ε. $\varepsilon = f(\Delta n)$.

THE STUDY OF THE INFLUENCE OF ELEMENTARY PARAMETERS OF HOMOGENIZATION PROCESS ON THE HOMOGENIZATION OF FAT PHASE GRANULES OF MILK AND CREAM

H. Popko, R. Popko, H. Komsta, A. Farian
Technical University, Lublin 1, skr. 189, Poland

The studies have shown that apart from the pressure, temperature etc., also the homogenizing valve construction has a considerable influence on the homogenization of fat phase.

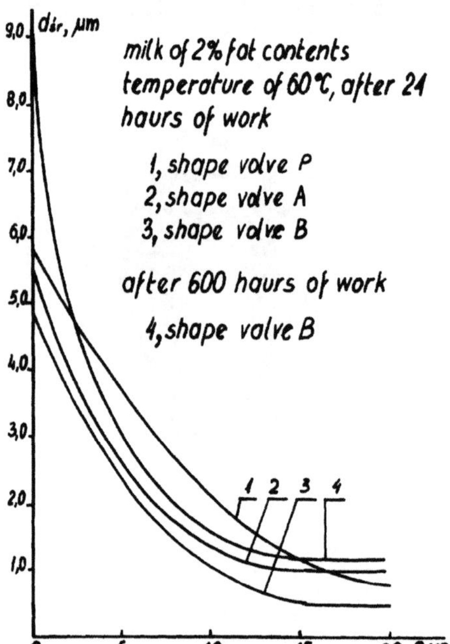

Fig.1.

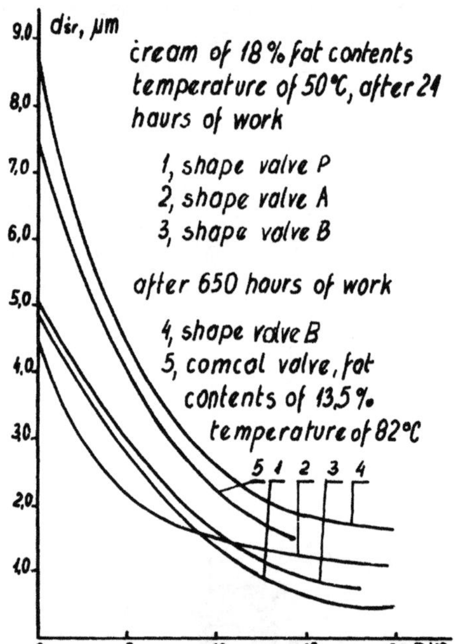

Fig.2.

Fig. 1. The dependence of medium diameter of fat granules on the valve type and the pressure during homogenization process of milk.

Fig. 2. The dependence of medium diameter of fat granules on the valve type and pressure during homogenization process of cream.

EFFECT OF CONCENTRATION OF SOME MILK CONSTITUENTS ON HEAT STABILITY

INTRODUCTION

The heat stability of concentrated milk is only poorly related to that of the non-concentrated milk. In this poster we describe some heat stability experiments in which the concentration of the main constituents of milk was varied separately. This may provide more insight into the role of each constituent.

EXPERIMENTAL PROCEDURE

Skimmilk (9 % solids-non-fat) was made from low-heat skimmilk powder. The concentration of lactose was varied by adding lactose to the milk, that of protein by ultra-filtration, that of salts by dissolving the milk powder into a simulated milk salt solution ('Jenness & Koops buffer'), that of fat by emulsifying anhydrous milk fat into skim-milk. The pH was adjusted by adding 0.5 N HCl or 0.5 N NaOH. Stainless steel tubes (content 3.8 ml) filled with 3.5 ml milk were rotated in an oil bath at 130 °C. The heat coagulation time (HCT) was determined as the time needed to cause visible coagulation. The samples were further analyzed after heating (e.g. for pH).

RESULTS

Effect of lactose: see Figure 1. The increased rate of coagulation outside the region of the relative minimum and maximum (pH 6.6 - 7.0) could be fully explained by the increased rate of acid production from lactose.

Effect of fat: see Figure 1. The homogenized fat globules behave as large casein micelles as the membranes consist mainly of casein; probably the casein adsorbed contains less κ-casein at the outside so that it coagulates more easily.

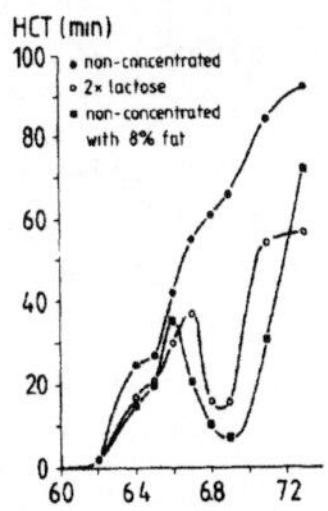

Fig. 1. Effect of lactose and fat

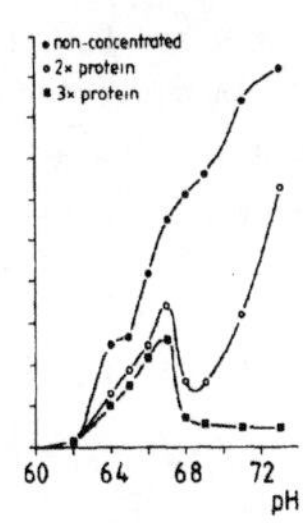

Fig. 2. Effect of protein

Effect of protein: see Figure 2. Processes involved are:

- increased concentration of coagulatable material (casein micelles)
- increased rate at which the pH is lowered because of hydrolysis of phosphate esters of casein
- increased concentration of serum proteins, of which especially β-lactoglobulin is known to decrease the heat stability.

Effect of salts: see Figure 3. Calcium and phosphate appear to have the most effect, probably because of Ca^{2+} induced coagulation and increased rate of precipitation of calcium phosphate. The increase in heat stability at pH < 6.7 may be due to a buffering effect of citrate.

Combined effects: see Figure 4.

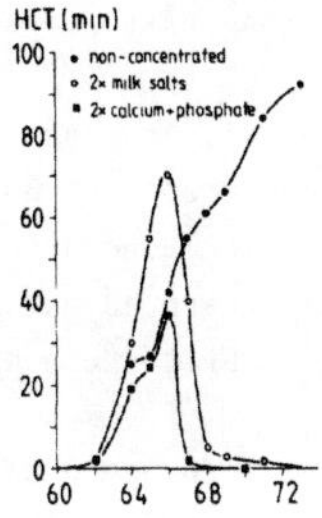

Fig. 3. Effect of salts

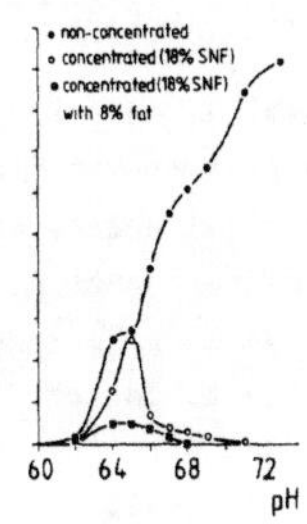

Fig. 4. Combined effects

CONCLUSION

The poor heat stability of concentrated milk is due to the concentration of lactose, protein and milk salts; milk salts appear to have the largest influence, especially at high pH. These effects are reinforced when homogenized fat globules are present as they behave as large casein micelles.

lh
Agricultural University
Department of Food Science
Laboratory of Dairying and Food Physics
Biotechnion, De Dreijen 12
6703 BC Wageningen, The Netherlands

M.A.J.S van Boekel
G.J. Rientjes
P. Walstra

QUALITY OF FROZEN STORED TALEGGIO CHEESE

ROBERTO GIANGIACOMO
ISTITUTO SPERIMENTALE LATTIERO CASEARIO
LODI ITALY

Taleggio cheese,a soft cheese typical of North-Italian dairy tradition, suffers seasonal market fluctuations. Its relatively short ripening period (30-40 days) does not allow long storage, therefore its suitability to freezing and long period frozen storage has been investigated.

The research has been carried out on ripe, whole cheese moulds.

Moulds have been frozen, after packaging under vacuum in Cryovac bags, in tunnel with forced air circulation at -40°C. After freezing the cheeses have been stored at -20°C for six months.

Chemical, physico-chemical, and organoleptic characteristics have been determined on the ripe fresh, 3 months frozen and 6 months frozen moulds.

Chemical determinations included also gel electrophoresis and GLC of volatile FFA. Colour measurements have been carried out by Hunter tristimulus colorimeter. Mechanical properties have been measured by texture profile analysis at 60% compression using an Instron texturometer.

Organoleptic tests have been performed by 10 panelists, expert judges of the Taleggio cheese quality. The scores have been evaluated by variance analysis and Duncan test (P 0.05).

The results of chemical analysis, electrophoresis and GLC show a non significant progress of the proteolysis and lipolysis during a long storage period.

The slight increase in butyric acid indicates a very slow progress of lipolysis that however does not alter markedly the acceptability by the panelists.

From those data it can be stressed that freezing does not considerably alter the Taleggio cheese, allow its storage up to 6 months with no substantial modifications in the chemical, physico-chemical, and organoleptic characteristics. This technique could be suitable, therefore, also for quality cheeses.

OPTIMIZING THE BUTTER CHURNING PROCESS

N. OLESEN, N. ANDERSSON and H. DANMARK

The Danish Government Research Institute for Dairy Industry, Roskildevej 56, DK-3400 Hillerød, Denmark

INTRODUCTION

Background

In continuous butter machines the churning section is the most energy consuming and often the capacity limiting part.

Aim

Developing a method of "prechurning" of the cream, ahead of the proper churning in the butter machine, is achieve

1. a reduction of the total energy consumption
2. an increase of the capacity of the butter machine

Scope

This poster deals with the introductory investigations carried out in laboratory scale.

MATERIALS AND METHODS

Sweet cream used in the experiments.

Under standardized conditions the necessary time for churning 400 g samples of cream was measured, and the effect of the prechurning treatment was expressed by the reduction of this time of churning related to that of untreated cream.

Three methods of prechurning were investigated.

A. Ultrasonic Treatment

The cream was pumped through a small chamber mounted on the horn of an ultrasonic transducer.

Volume of the chamber	0.30 ml
Height over the horn	1.50 mm
Ultrasonic power	0 - 60 W
Ultrasonic frequency	26.8 kHz

B. Air Injection into the Cream

The cream was pumped through a pipe where a controlled amount of air was injected through a diffusor.

C. Ultrasonics and Air Injection

The cream was continuously subjected to first ultrasonic treatment as in A and thereafter air injection as in B. The arrangement is shown in figure 1.

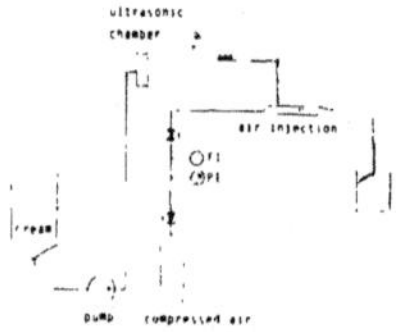

Figure 1. Arrangement for prechurning by ultrasonic treatment and air injection.

RESULTS

A. Ultrasonic Treatment

As operating power 20, 40 and 60 W were used. Cream treated at 0 W was used as the reference. As shown in figure 2 the reduction of time of churning was significant at all levels and the greatest reduction was 10.9% at 60 W.
By changing the flow rate of the cream the effect of the duration of the ultrasonic treatment was investigated. As shown in figure 3 the tested times of treatment did not yield significantly different prechurning effects.

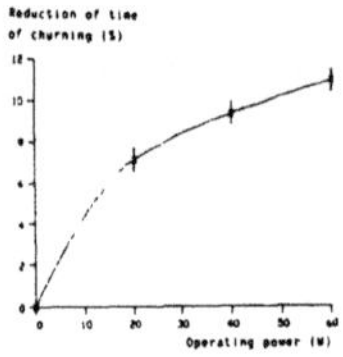

Figure 2. Ultrasonic treatment.
Reduction of time of churning versus operating power.

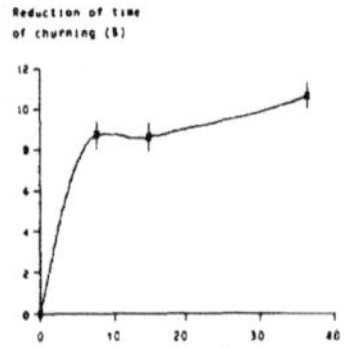

Figure 3. Ultrasonic treatment.
Reduction of the time of churning versus duration of the treatment.

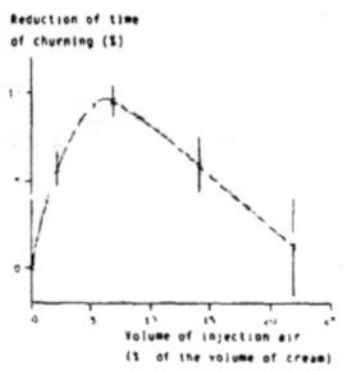

Figure 4. Air injection into the cream.
Reduction of time of churing versus
amount of air injected.

B. Air Injection into the Cream

Air volumes of 2, 7 14 and 22% related to volume of cream were investigated. Cream treated with 0 % air was used as the reference. As shown in figure 4 the greatest reduction was 9.5 % at an amount of 7 % air. Cream with 22 % air was not significantly better than the reference.

C. Ultrasonics and Air Injection

Combinations of 40 and 60 W operating power and of 4, 7 and 10 % air were investigated. The 0 - 0 combination was used as the reference. The reduction of time of churning as not greated than using ultrasonics or air injection separately.

REFERENCES

256th report from The Danish Goverment Research Institute for Dairy Industry, 1984.

King N.: The Theory of Churning.
Dairy Science Abstracts, 15, 590 - 606, 1953

MUlder H. & Schols J.J.G.: Churning in the absence of foam.
XIII International Dairy Congress, 706 - 709, 1953

Walstra P & Jenness R.: Dairy chemistry and Physics, John Wiley & Sons, Inc., 1984

MEMBRANE FOULING IN ULTRAFILTRATION OF DAIRY LIQUIDS*

Th. van den Boomgaard, T. Robbertsen, W. Versluis, J.H. Hanemaaijer
Netherlands Institute for Dairy Research, P.O. Box 20, 6710 BA EDE, The Netherlands

Membrane fouling is a serious limitation of membrane processes. In order to reduce membrane fouling insight into adsorption phenomena has to be obtained. For dairy liquids the most abundant foulant is protein. Protein adsorption onto membranes has been studied revealing the necessity of the knowledge of the specific surface area of membranes. Depending on the pore size protein can adsorb onto the inner membrane surface. Fouling can be reduced by choosing membranes with small pores.

MONDAY – POSTER 21

EXTRACTION OF MILKFAT WITH SUPERCRITICAL CARBONDIOXYDE*

J.E. Schaap, J. Straatsma and J.T.M. Escher
Netherlands Institute for Dairy Research, P.O. Box 20, 6710 BA EDE, The Netherlands
Presented by Dr. Jr. H.T. Badings, NIDR.

Milk fat was continuously extracted with supercritical carbondioxyde. Yields and composition of the fractions were determined at various temperatures and pressures. The phase diagram of milk fat-CO_2, the selectivity for the triglycerides of milk fat and the amount of free fatty acids in extract and residue were estimated at pressures upto 500 bar. It is concluded that extraction opens a possibility for additional use of milk fat.

*Complete poster available from authors.

NIZO

The effect of protein association on rejection and osmotic pressure during ultrafiltration

G. B. van den Berg[1], J. H. Hanemaaijer[1], C. A. Smolders[2].

Introduction

During ultrafiltration of protein containing liquids usually a considerable flux decline is observed. Besides by solute adsorption, pore-blocking, etc. the permeate flux is limited by the osmotic pressure difference across the membrane, which is further increased by concentration polarization (solute build-up at the membrane surface) and hence also influenced by the rejection of the solute.
For the whey protein β-lactoglobulin its concentration dependent association behaviour can be expected to influence both the rejection and the osmotic pressure build-up because of the different solute characteristics of the monomers and the dimers.

The monomer-dimer equilibrium of β-lactoglobulin

Several genetic variants of β-lactoglobulin exist each with slightly different properties. The protein tends to form oligomers, mostly dimers and some octamers, depending on pH, temperature and concentration. From literature data:

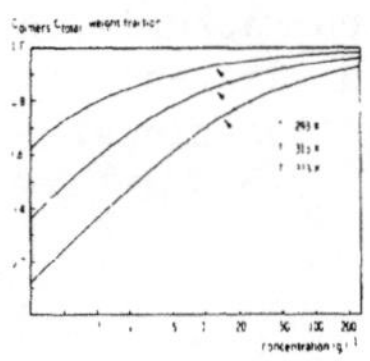

The fraction of dimers of β-lactoglobulin B as a function of temperature and concentration (pH = 7.0 and I = 0.1 M)

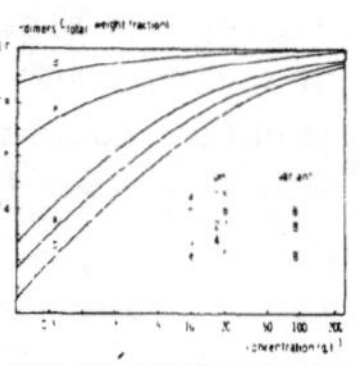

The fraction of dimers of β-lactoglobulin B and C as a function of pH and concentration (T = 293 K and I = 0.1 M)

The effect of the association of β-lactoglobulin on rejection

From several independent ultrafiltration experiments rejection data were gathered. The rejection $R = 1 - (C_p/C_b)$ is plotted as a function of the concentration in the retentate.

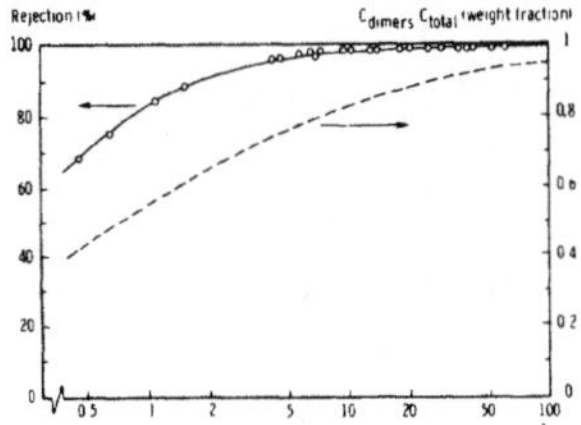

Full curve: the rejection of β-lactoglobulin as a function of retentate concentration (Iris 3038 membrane, pH = 6.6, T = 323 K and I = 0.1 M)
Dashed curve: fraction of dimers for the same conditions (K_{eq} = 2.90 10^{-5} mol l⁻¹)

A rejection of 100% for dimers and 50% for monomers could agree with the measured rejection at low concentrations. At higher concentrations the monomer rejection increases which can be explained by the increasing influence of additional ultrafiltration phenomena such as protein adsorption and pore-blocking.

The osmotic pressure of β-lactoglobulin solutions

The osmotic pressure can be calculated starting from the basic thermodynamic equation for the osmotic value of non-ideal solutions:

$$\Pi = \frac{RT}{M}\left(C + B_2C^2 + B_3C^3 + \ldots\right)$$

The osmotic pressure is calculated by taking into account the ideal Donnan-effects and the excluded volume (according to Vilker et al.) and using the protein association data.

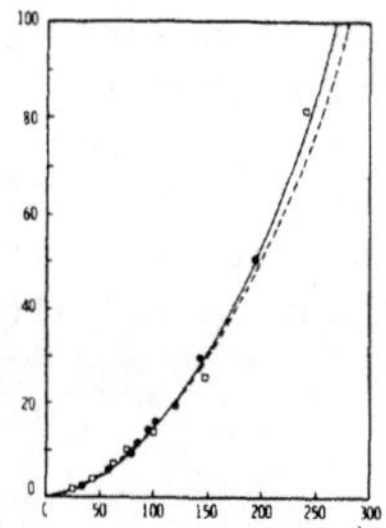

The osmotic pressure of β-lactoglobulin as a function of concentration (pH = 6.6, T = 323 K and I = 0.1 M)
Full curve: measured osmotic pressure; the membranes used were □ Amicon Diaflo UM 30 and • Abcor HFK 131
Dashed curve: calculated osmotic pressure

High concentrations, as they can be reached at the membrane surface during ultrafiltration, can result in rather high osmotic pressure differences, e.g. Π = 85 kPa at 250 g l⁻¹ and Π = 260 kPa at 400 g l⁻¹. These osmotic pressures will reduce the driving force ($\Delta P - \Delta\Pi$) considerably, resulting in low permeate fluxes.

The reduced osmotic pressure

Information about the state of association of β-lactoglobulin can be obtained by plotting the reduced osmotic pressure (Π/c) against the concentration since van 't Hoff's law reads

$$\frac{\Pi}{C}\bigg|_{\lim C \to 0} = \frac{RT}{M}$$

For β-lactoglobulin Π/c = 146.7 Pa l g⁻¹ for monomers (with M = 18,300) and Π/c = 73.3 Pa l g⁻¹ for dimers (with M = 36,600).
From the figure, where Π/c = 79 Pa l g⁻¹ by extrapolation, it can be concluded that β-lactoglobulin mainly consists of dimers.

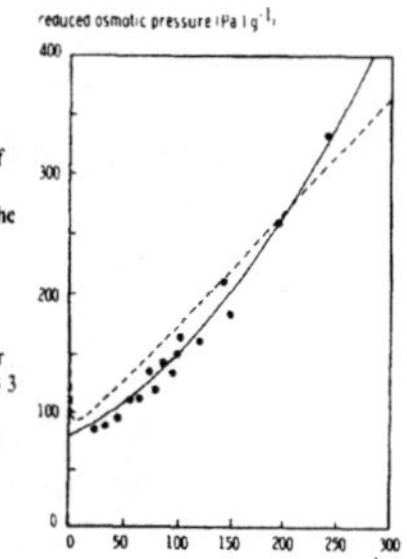

The reduced osmotic pressure of β-lactoglobulin as a function of concentration (pH = 6.6, T = 323 K, I = 0.1 M)
Full curve: experimental values. Dashed curve: theoretical values

The influence of pH on osmotic pressure

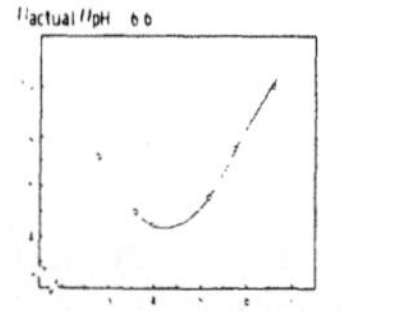

The osmotic pressure of β-lactoglobulin as a function of pH (concentration = 100 g l⁻¹, T = 323 K and I = 0.1 M)

A comparison is given for osmotic pressures measured at various pH-values with the osmotic pressure at pH = 6.6. It should be noticed that the minimum around pH = 4.5 is not the locus of the iso-electric point, which is at pH = 5.2. However, this minimum corresponds with the minimum in free enthalpy of the association reaction.

Conclusions

The state of association of β-lactoglobulin influences both the protein rejection and the osmotic pressure:
- the increasing rejection with concentration appears to be related to an increasing degree of protein association, although not exclusively
- the osmotic pressure can be described rather well up to high concentrations, using literature data on the equilibrium constant and other molecular properties
- the osmotic pressure of β-lactoglobulin at various pH-values reaches a minimum around pH = 4.5, where a minimum in free enthalpy of the association reaction exists.

[1]) Netherlands Institute for Dairy Research (NIZO), P.O. Box 20, 6710 BA Ede, the Netherlands.

[2]) Twente University of Technology, Dept. of Chemical Engineering, P.O. Box 217, 7500 AE Enschede, the Netherlands, which is the present address of the first author.

THE KLARO-GRAPH, AN AUTOMATIC DEVICE FOR MEASURING VISCOSITY, DENSITY AND HEAT STABILITY OF MILK (CONCENTRATES) AT TEMPERATURES UP TO 140°C

J.N. de Wit and G. Klarenbeek
Netherlands Institute for Dairy Research, P.O. Box 20,
6710 BA EDE, The Netherlands

The Klaro-graph is a falling ball viscosimeter, suitable for measurements of milk (concentrates) under sterilizing conditions up to 140°C. The operating principle is shown and some results are described with respect to industrial sterilization processes. Viscosity and density data, measured up to 140°C are related to flow characteristics in sterilizers, and heat stability data are discussed in reference to product properties during and after continuous flow sterilizations.

MONDAY – POSTER 24

THE (ULTRA)-PASTEURISATION OF WHIPPING CREAM: EFFECT ON CREAM PLUG FORMATION

A.C.M. van Hooydonk and A. Streuper
Netherlands Institute for Dairy Research, P.O. Box 20,
6710 BA EDE, The Netherlands

Cooling rate is the determining factor in formation of a solid cream plug during storage of in-line pasteurized cream. If coalescence occurs during heating, it plays a role as well. The latter effect is minimized by keeping the pressure high enough to prevent boiling and formation of air bubbles. The influence of cooling is explained by the combined effects of fatcrystallisation and shear forces.

*Complete poster available from authors.

HEAT TREATMENT OF WHIPPING CREAM: FOULING OF THE STERILIZATION EQUIPMENT *

J. Hiddink, R. Maas, M. Lalande*, A. Streuper, A.C.M. van Hooydonk
Netherlands Institute for Dairy Research, P.O. Box 20, 6710 BA EDE, The Netherlands, *INRA-Laboratoire de Génie Industriel Alimentaire, Villeneuve d'Ascq, France

Whipping cream was heat treated in a plate heat exchanger at about 120°C. The main fouling components of whipping cream appeared to be protein and ash. Fat did not greatly contribute to fouling. The amount of ash in the deposit increased with the temperature level. The correspondence between the calculated denaturation rate of whey protein, and the experimental rate of deposition of potein supports the hypothesis that thermal denaturation of whey protein plays a key role in the fouling process.

*Complete poster available from authors.

The Present Standard of Butter Processing and Buttermaking Equipment in Czechoslovakia

Forman, L., Dairy research institute in Prague
Nevečeřal, J., Machinery Chotěboř

Czechoslovakia is a country with a relatively high butter consumption wich is completely covered by the dynamically developped home butter manufacture. Therefore a great attention is paid to buttermaking respecting the field of processing as well as the development of equipment.
The butter is manufactured mostly of sweet cream. The majority of butter (about 80%) is marketed after cooling within 12 days after the manufacture as butter. Only continuous buttermaking is applied there by means buttermakers having the hour capacities 1,2,3 and 4 tons.
The Czechoslovak industry of food engineering produces all items of equipment for buttermaking lines with capacities 1 till 4 tons.h^{-1} except of cream separators and automatic butter packaging machines. One of the headlights of the inovation in the field of buttermaking equipment are the pasteurizers of the new generation for treatment of cream, They use hot water as heating medium and they are furnished with a suitable section of deodorization. These pasteurizers may be used for the first steps of modified Alnarp process of the typical ripening of summer and winter cream as well as for the one-step treatment of cream.
Further important innovation in the field of buttermaking is the new range of continuous buttermakers of ZKU-standard following the well tested buttermakers of KM-standard. Their commercial flexibility, processing reliability and possibility to manufacture butter with water content up to 42,5% are especially appreciated.
The continuous buttermakers of ZKU-standard consist of churning cylinder, separating cylinder, working section I, vacuum chamber and working section II.
Improved construction, namely perfected continuously controled driving speed, gives large flexibility in capacity and composition of butter. Also the noice level is reasonably low. New buttermakers can wash the butter grain and thus reduce the solid non fat content of butter by 0,1 to 0,4 %. In vacuum chamber there is possible to reduce air content of butter as low as to 1 v %. Relatively high water content of butter is suitable for production of modified kinds butter.

Study on the Butter Consistency

Forman, L., Dairy research institute in Prague
Matoušková, E., -"-
Pokorný, J., Institute of Chemical Technology, Prague
Štern, P., Institute of Hydrodynamics of Czechoslovak Akademy of Science, Prague

Consistency of butter was studied a) by penetration (cone $\lambda = 20^{\circ}$, W=102,0g; $\lambda = 40^{\circ}$, W=196,6g), b) by parameters of rotational rheometry (static yield value, dynamic yield value and apparent viskosity /1/). Cone and plate viscometer Feranti Shirley (Feranti Ltd, England) c) by the method of sensory texture profiling in the temperature range of 4-25 °C. The profile consisted of 8 descriptors, and unstructured graphical seales 150 mm long were used for rating. The textural characteristics were determined by picking, scratching, and cutting the sample with a knife, and by spreading on a slice of rye bread. The sensory panels consisted of 6-8 trained assessors, and the results were based on 40-80 responses.The repeatability was about 5-10% of the scale at the determination of various characteristics, respectively. The reproduceability, was about 10-25%, depending on the characteristic and on the temperature.
Values of penetration were transformed for "Haightons' yield value - C"/2/. Comparisson of Haightons' yield value (C) calculated from penetration and static yield value ($\tilde{\tau}_S$) determined by objective rheometric method showes that there exists exponential dependence between both parameters characterizided by equation $C = 40\,\tilde{\tau}_S^{\,0,7}$ with the value of the correlation coeficient r = 0,788.
It is interesting that between both mentioned parameters of margarine the depandance exixts of the same form $c = \tilde{\tau}_S^{\,0,9}$ only in absolute values of the coefficient and exponent.
The hardness by picking, scratching, and cutting, and on spreading on a slice of bread decreased with increasing temperature following semilogaritmical relationship which was close to linear. The effect of temperature on the spreadability, plasticity, and greasiness was more complicated passing through a flat maximum at higher test temperatures, and through greater deformations at each end of the scale. The regression between the hardness determined by sensory analysis (by cutting and picking) and both between the apparent viscosity and the static yield value were double logarithmic with relatively high correlation coeficients (r=0.704 and 0.857 respectively, N=33) but the rheological characteristics were not simply correlated with the spreadability and greasiness.

Verweilzeitverhalten in Plattenwärmeaustauschern

Chr. Kiesner und H. Reuter
Institut für Verfahrenstechnik der Bundesanstalt für Milchforschung in D-2300 Kiel (W.Germany)

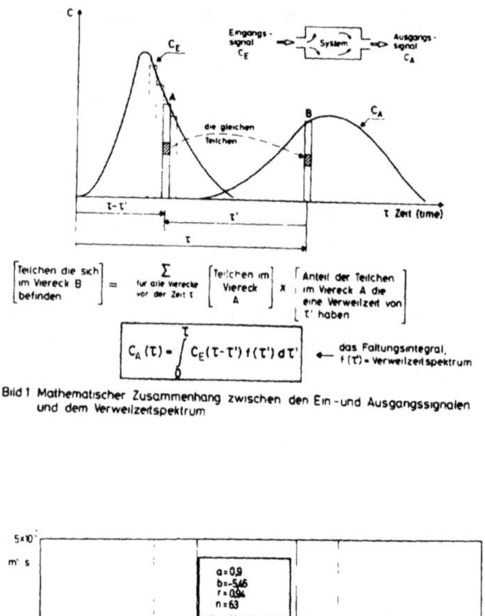

Bild 1 Mathematischer Zusammenhang zwischen den Ein- und Ausgangssignalen und dem Verweilzeitspektrum

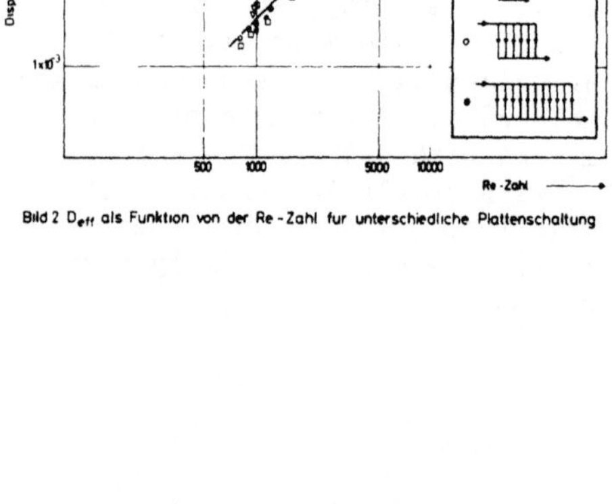

Bild 2 D_{eff} als Funktion von der Re-Zahl für unterschiedliche Plattenschaltung

Für die Ermittlung der Verweilzeitfunktion in Plattenwärmeaustauschern wurde die Methode des nichtidealen Markierungssignals und das Zwei-Meßstellensystem gewählt. Die Verknüpfung zwischen dem Eingangssignal und dem daraus folgenden Ausgangssignal kann mit Hilfe des Faltungsintegrals hergestellt werden, was Bild 1 wiedergibt. Für die Beschreibung des Verweilzeitspektrums wurde das Dispersionsmodell verwendet. Die Ergebnisse wurden als Funktion des Dispersionskoeffizienten von der Re-Zahl dargestellt. Das Bild 2 gibt beispielsweise den Verlauf der Funktion $D_{eff} = f(Re)$ für unterschiedliche Anzahl der parallelen Fließwege wieder. Die gefundene Funktion weist unabhängig von der Anzahl der parallelen Fließwege im untersuchten Re-Bereich einen linearen Verlauf auf.

INFLUENCE OF IONIC STRENGTH ON β-LACTOGLOBULIN PERMEABILITY OF POLYSULPHONE ULTRAFILTRATION MEMBRANES.

L. DE VALCK and A. HUYGHEBAERT
Laboratory of Food Chemistry and -microbiology
Faculty of Agricultural Sciences
Coupure links, 653 B-9000 GHENT Belgium

Molecules with a molecular weight higher than the cut-off value of the membrane should be retained in the concentrate, retention of 100% is not always realized. The retention of a membrane for a solute is a defined as R=1-Pm; Pm being membrane permeability. The retention of a molecule is influenced by its configuration (globular, unfolded or fibrillar). The concentration of β-Lg in the permeate (cut-off 50000) during ultrafiltration is as follows.

% volume reduction	β-Lg (μg/ml) synthetic solution (150 μS)	β-Lg (μg/ml) raw whey (4600 μS)
50	6.1	282
85	4.0	454
95	7.3	875

The evolution of the concentration of β-Lg and α-La in raw whey permeate (in μg/ml) during ultrafiltration over a membrane with cut-off value 20000 is given below.

% VR	45	50	70	75	85	95	98
α-La	35	40.4	51.6	56.5	32.7	51.2	115
β-Lg	6.6	7.1	15.6	21.8	22.7	50.7	144

Through membranes with cut-off 20000 no permeability of β-Lg was detected when whey was ultrafiltrated on a membrane (cut-off 20000) the permeation of β-Lg starts at a volume reduction of 50%. No permeability is observed for the synthetic solution even after a volume reduction of 95%. A second proteose peptone peak appears in the permeate of the ultrafiltrated synthetic solution on both membranes. The average weight of this compound is 3000 Dalton. From these data it is evident that β-Lg in a low ionic environment does not permeate through the membranes. This observation can be explained by the unfolding of the β-Lg molecule, so that the structural configuration is too large to permeate through the pores of the membrane. Enrichment of particular proteins can be obtained by UF under appropriate conditions.

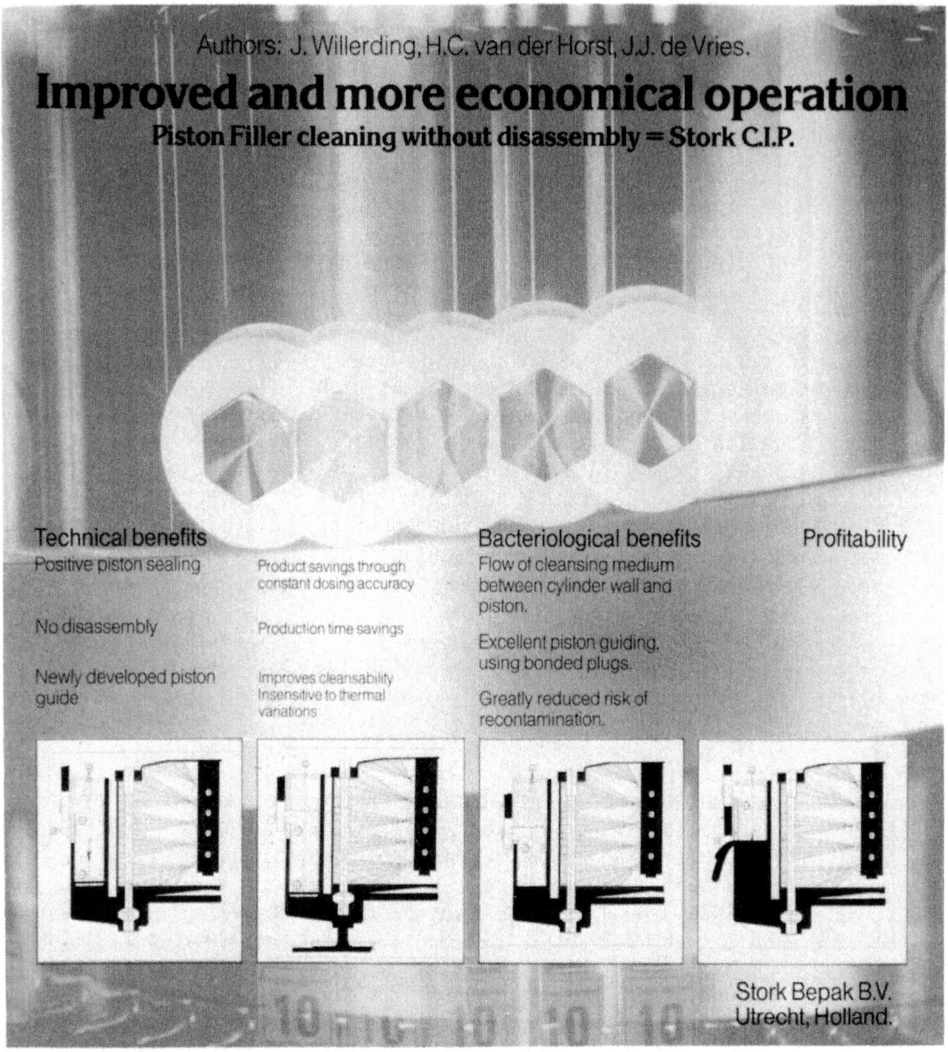
Authors: J. Willerding, H.C. van der Horst, J.J. de Vries.
Improved and more economical operation
Piston Filler cleaning without disassembly = Stork C.I.P.
Technical benefits
Positive piston sealing
Product savings through constant dosing accuracy
No disassembly
Production time savings
Newly developed piston guide
Improves cleansability Insensitive to thermal variations
Bacteriological benefits
Flow of cleansing medium between cylinder wall and piston.
Excellent piston guiding, using bonded plugs.
Greatly reduced risk of recontamination.
Profitability
Stork Bepak B.V.
Utrecht, Holland.

MODERN TRENDS IN UHT PROCESS-CONTROL

Ir. W. F. Hermans and Ir. A. Meijer*

Stork Central Research and Development Laboratories,
P.O. Box 103, 2150 AC Nieuw Vennep, The Netherlands

*Stork Amsterdam, P.O. Box 3007, 1003 AA Amsterdam,
The Netherlands

Variable production conditions of high capacity UHT-installations often occur due to short stops of one or more connected filling machines. Its influence on the processing values are minimized by variable active heatexchanging surfaces. Applying micro-computer technology enables to control the process on F_o-, B- and C-values instead of on temperature, hence resulting into more constant product quality regardless variations in production conditions.

MURUHUA NAIFEN----A KIND OF HUMANIZED MILK POWDER IN CHINA

By JIN SHI-LIN, LI WEI-JIANG, SUN JIA-SHAN

Inner Mongolia Light Industry Scientific Research
Institute, 30 East Hairlar Road Huhhot
Inner Mongolia P.R. China

SUMMARY. Breast-feeding is on the decline in China. From 1950 to 1980 the percentage of breast-feeding Beijing mothers down from 81% to 22% . Only 24.8% of Shanghai mothers were breast-feeding in 1980 . Therefore some kinds of infant formula is necessary. MURUHUA NAIFEN is a kind of humanized milk powder which is alter the protein of cow's milk in its casein and lactalbumin for a 40:60 ratio of there two components by adding demineralized whey by electrodialysis. Vegetable fat is added to provide sufficient polyunsaturated fatty acids. The Ca/P ratio is regulated and enriched with vitamins and iron-fortified.

MURUHUA NAIFEN the infant formula is milk and whey-based formulation, use the demineralized whey as source of protein and lactose. Again vegetable oils are used as source of fat and polyunsaturated fatty acids enriched with vitamins and iron-fortified.

Gross nutrient composition of MURUHUA NAIFEN

The infant formula was designed as Table 1.

Table 1. Gross nutrient composition of MURUHUA NAIFEN

Composition	content (per 100g powder)
Fat (g)	24-30
Protein (g)	12-15
Casein:Lactalbumin	40:60
Lactose (g)	52-58
Minerals (g)	< 3.2
Moisture (g)	< 3
Vitamin A(IU)	2000
B_1 (mg)	0.5
C (mg)	50
D (IU)	350
E (mg)	9
Folic acid (mg)	0.3
Linolic acid (mg)	1500
Ca (mg)	320
P (mg)	270
Fe (mg)	6
Energy (KJ)	2051-2219

Processing MURUHUA NAIFEN powdered infant formula

Dried formula may be prepared either by blending dry ingredients or by drying a mixture of liquid ingredients. MURUHUA NAIFEN is the latter type.

Ingredients:

Fresh milk (filtered and deaerated)
Demineralized whey (by electrodialysis)
Fresh cream
Vegetable oil (refined corn oil)
Vitamin A----Retinyl acetate or Retinyl palmitate
D----D_2 or D_3
E----d-α-tocopherol
B_1 ---Thiamin hydrochloride
B_2 ---Riboflavin
B_6 ---Pyridoxine hydrochloride
B_{12} --Cyanocobalamin
C----Ascorbic acid
Folic acid
Nicotinic acid or Nicotinamide
iron----Ferrous Sulfate

All are accord with Codex standards of China pharmacopoeia

Saturated fatty acids and unsaturated fatty acids content of refined corn oil from SHIJIAZHANG North China pharmaceutical manufactory as Table 2.

Table 2. Saturated and unsaturated fatty acids content in corn oil

Saturated fatty acids	content (%)
C_{14}	0.1-1.7
C_{16}	8-12
C_{18}	2.5-4.5
Unsaturated fatty acids	
$C_{18:1}$	19-49
$C_{18:2}$	34-62
$C_{18:3}$	0-2.9

Technical processing

See Fig.1 for a flow diagram of the process for preparing dry infant formula MURUHUA NAIFEN.

```
                                 Cream
                                   ↓
                             ------------
Fat Soluble Vitamins →      | Mixing     |   ← Vegetable Oil
                             ------------
                                   ↓
                          ----------------
                          | Homogenation  |  ← Milk
                          ----------------
                                   ↓
        Water Soluble        ------------
        Vitamins and    →   |  Minxing  |   ← Demineralized whey
        Ferrous Sulfate      ------------
                                   ↓
                          ----------------
                          | Pasteurization |
                          ----------------
                                   ↓
                          ----------------
                          |  Evaporation   |
                          ----------------
```

```
        ↓
 -----------------
| Spray drying    |
 -----------------
        ↓
 -----------------
|   Packaging     |
 -----------------
```

Fig.1 Flow diagram of the process for MURUHUA NAIFEN

Major Composition and Nutrient of the MURUHUA NAIFEN commercial products as a result of analysis shown in Table 3.

Table 3. Major Composition of the MURUHUA NAIFEN

Composition Content	1	2		1	2
			Fatty acids(g/100g total acids)		
Protein (g)	16.4	14.8			
Fat (g)	25.8	29.9	16:0	33.7	
Lactose (g)	50.7	52.0	18:0	7.01	
Ash (g)	2.8	2.7	18:1	24.09	
Ca (mg)	564	630	18:2	13.41	
P (mg)	328	420	18:3	2.38	
Fe (mg)	7.58	9	8:0	0.56	
Zn (mg)	5.14	3	10:0	1.88	
Vitamin A(Iu)	1200.6	1500	10:1	0.24	
B_1(mg)	0.65	0.50	12:1	2.72	
B_2(mg)	3.29	0.50	14:0	9.50	
C(mg)	46.4	50.0	14:1	2.19	
D(Iu)	263.5	400	15:0	1.60	
E(mg)	8	ND	15:1	0.30	
Cholesterol(mg)	61.9	ND	Amino acids (mg)		
Moisture (g)	1.5	2.5	Asp	1141.4	1287.0
K (mg)	420		Thr	592.0	674.6
Na(mg)	430		Ser	671.6	785.5
Mg(mg)	78		Glu	2783.9	3265.4
Mn(mg)	3.5		Pro	1101.3	1306.1
Cu(mg)	trace		Gly	263.4	310.9
Se(mg)	ND		Ala	522.8	
			Cys	241.4	230.0
			Val	804.4	918.4
			Met	319.6	373.2
			Ile	689.1	798.0
			Leu	1487.8	1724.0
			Tyr	614.9	718.8
			Phe	628.9	727.4
			His	380.2	337.1
			Lys	1150.2	1335.1
			Arg	424.4	503.5
			Trp	ND	ND

Animal experiment

MURUHUA NAIFEN the dried infant fomula commercial product through the animal experiment result as Table 4.

Table 4. Growth of rats

		1 Infant nutrition prowder*	2 S--26**	3 MURUHUA NAIFEN
Start age(week)		4	4	4
Finish age(week)		13	13	13
Feeding period(week)		9	9	9
Animal sex & number***		♂5+5♀	♂5+5♀	♂5+5♀
Feed consumption (g)		842±101	561±81.7	701±72.7
Body weight (g)	initial	43.2±3.5	43.32±3.8	43.31±3.8
	finish	249.8±44.2	151.5±30.5	203.5±44.8
	gain	206.6±44.9	108.1±31.2	160.2±44.5
Body weight gain per 100g feed (g)		24.3±2.68	19.21±3.96	22.56±4.58
Body length (cm)	initial	11.8±0.42	11.95±0.37	11.90±0.39
	finish	21.95±1.30	19.25±1.18	20.85±1.29
	gain	10.15±1.42	7.3±1.16	9.0±1.73
Hemoglobin (g/100mL)	initial	9.35±1.23	11.8±1.01	12.0±1.53
	finish	14.0±0.80	14.3±0.86	13.7±1.33

*By China Academy of Medical Science Hygienic Institute fomula designed
** SMA S-26 a product of Wyeth Laboratories U.S.A
*** Rats from China Academy of Medical Science Hygienic Institute

Infant feeding experiment

MURUHUA NAIFEN the powdered infant fomula by reconstitution with water to a 12.5% T.S. milk compared with Human milk the gross nutrient composition shown in Table 5.

Table 5. Gross nutrient composition per 100g

Nutrient composition		12.5%T.S.MURUHUA NAIFEN	Human milk
Protein (g)		1.85	1.5
Fat (g)		3.74	3.7
Lactose (g)		6.5	6.9
Ash (g)		0.33	0.3
Ca (mg)		79	34
P (mg)		52.5	15
Fe(mg)		1.13	0.1
Zn(mg)		0.735	ND
Vitamin	A(iu)	187.5	176.6
	D(iu)	50	0.4-10
	B_1 (μg)	62.5	16
	B_2 (μg)	62.5	43
	C (mg)	50	4.3
Amino acids(mg/L)			
	Thr	843.25	610
	Phe	909.25	470
	Val	1148	803

Table 5. Continuous

Leu	2155	910
Ile	997.5	710
His	421.38	280
Lys	1668.88	830
Met	466.5	140
Trp	ND	230

MURUHUA NAIFEN compared with Human milk feeding for 30 of under 2 month Age infant during 3 month separately the result showing in Table 6

Table 6. MURUHUA NAIFEN Compared with Human milk in Infant feeding

	Human milk			MURUHUA NAIFEN			t value test	P
	start	finish	gain	start	finish	gain		
Body weight (kg)	(pre-feeding)		(past-feeding)	(pre-feeding)		(past-feeding)		
mean $\bar{X}$	4.8900	7.4537	2.5637	4.5487	7.2047	2.6560	.5417	
SD	(3.00–7.70)	(5.45–11.0)	0.1375	(3.00–5.80)	(5.30–8.95)	0.1007		P>0.05
Body length (cm)								
mean $\bar{X}$	56.1067	64.8167	8.7100	54.5700	64.0700	9.5000	1.1252	
SD	(48.00–61.00)	(59.00–72.20)	0.4435	(45.00–60.00)	(59.00–70.00)	0.5443		P>0.05
Chest measurement (cm)							2.8465	
mean $\bar{X}$	37.4467	42.2167	4.7700	36.2200	42.3867	6.1667		
SD	(32.00–43.00)	(37.50–46.50)	0.3613	(30.00–42.00)	(40.00–46.60)	0.3320		P>0.05
Sitting height (cm)							0.1531	
mean $\bar{X}$	37.1867	42.8900	5.7033	36.2933	42.0867	5.7933		
SD	(10.00–1500)	(10.00–14.00)	0.2601	(7.00–15.50)	(10.00–13.00)	0.3704		P>0.05
Hemogloim (g/100ml)							1.7675	
mean $\bar{X}$	11.7667	11.7000	-0.0667	11.0667	11.8000	0.7333		
SD	(10.00–15.00)	(10.00–14.00)	0.2601	(7.00–15.50)	(10.00–13.00)	0.3704		P>0.05
Calcium Concentration in Blood (g/100ml)							0.2694	
mean $\bar{X}$	9.4667	10.8133	1.3467	9.4700	10.6933	1.2233		
SD	(8.00–12.00)	(8.50–13.00)	0.092	(8.00–12.00)	(5.90–13.00)	0.3526		P>0.05

Table 6 continuous

Phosphour Concentration in blood (g/100ml)							0.2635	
mean $\bar{X}$	4.9950	4.8933	-0.1017	5.4000	5.4017	0.0017		$P>0.05$
SD	(3.60–6.50)	(3.75–6.50)	0.1821	(3.50–12.00)	(3.75–10.00)	0.3473		
Alkaline Phosphatasé (mg/100ml)							0.7638	
mean $\bar{X}$	34.1367	24.9767	-9.1600	38.9333	26.7733	-12.1600		
SD	(18.50–93.00)	(10.40–44.00)	2.5693	(12.70–41.04)	(12.00–41.00)	2.9709		$P>0.05$

Survival and enzymic activity of Yersinia enterocolitica in refrigerated Ultra Heat Treated Milk.

Dr.R.N.Roy
Department of Applied & Life Sciences.,
The Queen's College.,
1.Park Drive.,
Glasgow.G3 6LP
United Kingdom.

Yersinia enterocolitica is a species of the genus Yersinia,family Enterobacteriacea.It has been isolated from a variety of food: beef, lamb,raw and pasteurized milk,shell fish and drinking water. Certain serotypes,3 and 9,are reported to be potentially enteropathogenic and a number of instances of infection have been reported.
An investigation into the prevalence of Yersinia enterocolitica in cold-stored raw milk and cottage cheese revealed them to be the predominant species surviving cold storage.It was anticipated that their rate of metabolic activity would progressively diminish,but the activity of their lipolytic and proteolytic extra- and intracellular enzymes may affect the organoleptic qualities of the dairy products. To test this hypothesis,an investigation was carried out to ascertain the survival of Yersinia enterocolitica inoculated into Ultra Heat Treated milk,stored at 4.C.,and ascertain qualitatively their proteolytic and lipolytic activities.
The results indicate that over a period of fourteen-day refrigerated storage,one strain of Yersinia enterocolitica (NCTC 10598) continued to profilterate in the U.H.T.milk,whereas another strain (NCTC 10460) reached a peak of growth on the eighth day and then declined.The pH of the former strain dropped from 6.5 to 5.6,but the latter dropped only to 6.1. There appeared little proteolytic activity,measured as zones of clearing on Casein Agar. The lipolytic activity showed marked increase,measured as zones of clearing on Tributyrin Agar.

The results of the investigation indicate that lipolytic activity of the extracellular and intracellular enzymes of Yersinia enterocolitica are probable in cold stored dairy products,even though their viable numbers continue to decline over the period of storage. The implications in terms of organoleptic qualities of refrigerated dairy products which may have been contaminated is a matter of concern in view of the enteropathogenic character of certain serotypes of Yersinia enterocolitica.

THE COMPOSITION OF THE FREE FATTY ACID FRACTION IN MILK, CREAM AND BUTTER

SVEN BORGSTRÖM AND HANS JÖNSSON
Swedish Dairies' Association (SMR),
Central Laboratory, Box 205, 201 22 Malmoe, Sweden

OBJECTIVE. To quantitatively follow the composition of the free fatty acid (FFA) fraction when going from milk to cream and cultured butter.

METHOD. Needs et al. Journal of Dairy Research (1983) 50 321-329.

RESULTS. The table shows that the composition of the FFA fraction and the composition of the triglyceride bound fatty acids in milk are similar with some minor variations.

The diagram illustrates that the share of butyric, caproic and caprylic acids is much lower in cream and butter than in milk and that of palmitic and oleic acids is correspondingly higher.

DISCUSSION. For the formation of rancid flavour the water-soluble short chain FFA are of prime interest as their contribution to this off-flavour is much greater than that of the long chain FFA.

In modern dairying the lipolysis problem can in most cases be traced back to the mechanical treatment of the milk on the farm. By mixing of air into the milk which occurs during milking and in the farm tank the indigenous milk lipase is activated. After the pasteurization this enzyme is inactivated and thus the lipolysis stopped. This fact, together with the results presented, show that the determination of the total acidity of butterfat is not enough to judge the rancid taste of a butter. Provided the lipolysis has occurred in the milk a butter can have an acid number of 1,8 meq/100 g fat without rancid taste. On the other hand a butter having an acid number of 1,2 meq/ 100 g fat can be rancid if the lipolysis is due to the presence of heat stabile bacteria lipases in the butter. The conclusion is that determination of the content of the lower FFA per se is a better way to judge a butter for rancidity than to measure the total acidity of the butter fat.

TABLE

THE COMPOSITION OF THE FREE AND TRIGLYCERIDE BOUND FATTY ACIDS IN MILK

Acid	Free fatty acid fraction %	Triclyceride bound fatty acids %
Butyric	4,1	4,7
Caproic	2,8	2,7
Caprylic	2,4	1,6
Capric	4,1	3,7
Lauric	5,0	4,5
Myristic	9,2	13,7
Palmitic	21,4	27,3
Stearic	10,7	9,6
Oleic	30,1	22,0
Linoleic	2,7	1,5
Linolenic	0,9	0,5

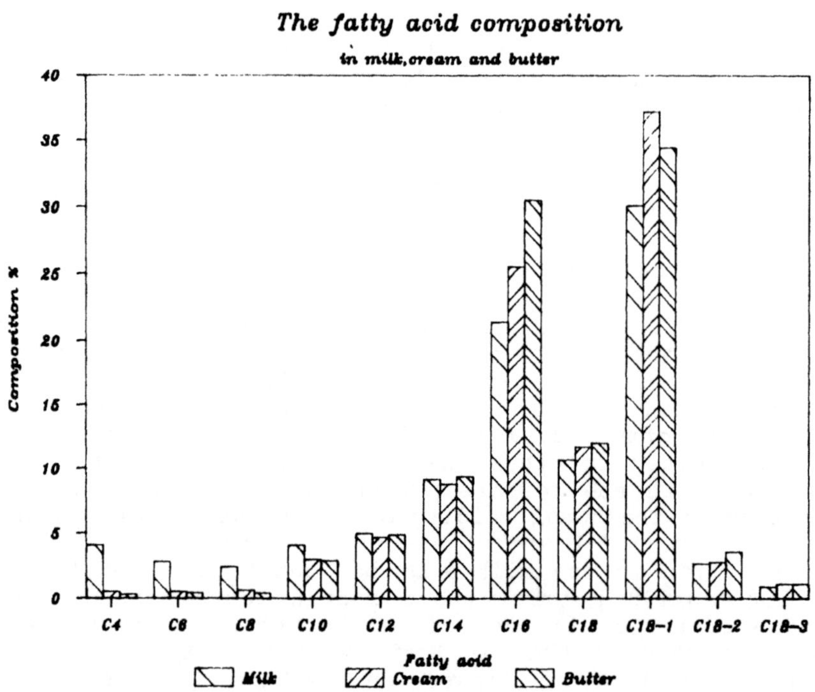

MILK PROTEIN POLYMORPHISM AND ITS EFFECT ON COMPOSITION AND TECHNOLOGICAL PARAMETERS

A.-M. BECH AND K. ROTVIG KRISTIANSEN

Introduction

In milk from three Danish cattle breeds, genetically controlled variations have been found in β-lactoglobulin and α_{s1}-, β-, and κ-casein (Larsen & Thymann (1966)). However no investigations concerning the correlation between genetic variation and yield or technological properties of the milk were carried out.

Such studies have recently been published from several laboratories (Mariani et al. (1976), McLean et al. (1984), Ng-Kwai-Hang et al. (1984), Sadler et al. (1968), Schaar et al.(1985)). The results indicate that genetic variants of the milk proteins have influence on technological properties measured as cheese yield, acid production, coagulum strength and heat stability.

The question arises as to whether the breeding of milk cattle towards higher milk and fat yield has at the same time selected animals with unfavourable genetic variants of the milk protein.

The purpose of this study is to elucidate whether such a selection has taken place.

Materials and methods

Milk from 552 cows from eight herds comprising 158 Jersey, 170 Red Danish Dairy Cattle (RDM), and 224 Black and White Danish Dairy Cattle (SDM) was phenotyped. The phenotypes were analyzed by means of isoelectric focusing (pH 2.5-7) in agarose gels containing 7 M urea.

305-days records for each of the 552 cows on milk, protein, and fat yield for the first and second lactation period have been collected. 178 cows were selected on the basis of their genetic variants, so that a 1-3-3-3 and a 2-4-2-3 (α_{s1}-, β-, κ-casein, β-lg) factorial analysis of variance could be carried out for RDM and SDM, and Jersey, respectively. By these statistical analyses it is possible to evaluate the main effects as well as 2-factor interactions of the genetic variants on technological properties.

Morning and evening milk samples are collected twice, with an interval of 2-3 months from each of the 178 cows. They are phenotyped and analysed for casein, whey protein, non protein nitrogen, acid production, heat stability, coagulation properties and syneresis.

Results and discussion

Genetic variation

The distribution of α_{s1}-, β-, and κ-casein phenotypes and the β-lactoglobulin phenotypes in populations of three Danish cattle breeds is given in table 1. Gene frequencies calculated from data in table 1 are given in table 2 as well as the results of a χ^2-test by which the gene frequencies have been compared with those found by Larsen and Thymann (1966).

In α_{s1}-casein the dominating component is the B-variant in all the three breeds, but also the C-variant is quite common in Jersey. In fact the C-variant has become much more frequent (0.29) than in the 1966-study (0.05). In β-casein the A^1-and A^2-variants dominate in RDM and SDM, while A^2-and B-variants are the most common in Jersey. A comparison with the results of Larsen & Thymann is difficult, because they did not determine the different A-variants. In κ-casein the A-variant is dominating the B-variant in RDM and SDM, while the opposite is the case in Jersey. Also here a significant shift has taken place (Jersey. κ-casein A 0.52 → 0.29 SDM κ-casein A 0.68 → 0.85) For β-lactoglobulin a change has taken place in Jersey only (β-lg A 0.42 → 0.30).

In case the genetic variants have any influence on the technological properties of the milk, such changes in variant composition certainly will influence the processing and the quality of milk products.

Table 1. The observed distribution of phenotypes among 552 cows.

Breed	No. of animals	α_{s1}-casein AB	BB	BC	BE	CC	β-casein A^1A^1	A^1A^2	A^1A^3	A^1B	A^2A^2	A^2A^3	A^2B	A^3B	BB	κ-casein AA	AB	BB	β-lactoglobulin AA	AB	AC	BB	BC
Jersey	158	-	73	79	-	6	6	10	-	6	47	1	74	-	14	11	71	76	14	67	1	75	1
RDM	170	1	167	1	-	-	85	50	-	18	12	-	3	-	1	108	50	11	-	37	-	132	-
SDM	224	-	208	15	1	-	54	117	11	9	26	2	4	1	-	165	49	10	65	113	-	46	-

Table 2

Gene frequencies. Danish dairy cattle 1985
(552 cows)
(compared with results of Larsen & Thymann (1966)
by a χ^2-test)

		Jersey (158 cows)	RDM (170 cows)	SDM (224 cows)
α_{s1}-casein	A	0	0.003	0
	B	0.712	0.994	0.964
	C	0.288	0.003	0.033
	E	0	0	0.002
χ^2-test		***	N.S	***
β-casein	A^1	0.089	0.706	0.547
	A^2	0.566	0.226	0.391
	A^3	0.003	0	0.031
	A (total)	0.658	0.932	0.969
	B	0.342	0.068	0.031
χ^2-test		N.S	N.S.	**
κ-casein	A	0.294	0.788	0.846
	B	0.706	0.212	0.154
χ^2-test		***	N.S	***
β-lactogl.	A	0.304	0.109	0.542
	B	0.690	0.891	0.458
	C	0.006	0	0
χ^2-test		**	N.S	N.S

** P <0.01 *** P <0.001

Influence on technological parameters

Milk sample collection and analyses are due to end in June 1986, and the results of this part of the investigation will be presented on the poster. Also the influence of genetic variation on milk, protein, and fat yield will be presented.

References

Larsen, B. and Thymann, M. (1966). Acta veterinaria Scandinavica 7, 189-205. Studies on milk protein polymorphism in Danish cattle and the interaction of the controlling genes.

Mariani, P., Losi, G., Russo, V., Castagnetti, G.B., Grazia, L., Morini, D. & Fossa, E. (1976). Scienza Tecn. Latt.-casear. 26, 437-484. Prove de caseificazione con latte caratterizzato dalle varianti A e B della -caseina nella produzione del formaggio Parmigiano-Reggiano.

McLean, D.M., Graham, E.R.B., Ponzoni, R.W. and McKenzie, H.A. (1984). Journal of Dairy Research 51, 531-546. Effects of milk protein genetic variants on milk yield and composition.

Ng-Kwai-Hang, K.F., Hayes, J.F., Mosley, J.E. and Moxley, H.G. (1984). Journal of Dairy Science 67, 835-840. Association of Genetic Variants of Casein and Milk Serum Proteins with Milk, Fat, and Protein Production by Dairy Cattle.

Sadler, A.M., Kiddy, C.A., McCann, R.E. and Mattingly, W.A. (1968). Journal of Dairy Science 51, 28-30. Acid Production and Curd Toughness in Milks of Different α_{s1}-Casein Types.

Schaar, J., Hansson, B. and Pettersson, H.-E. (1985). Journal of Dairy Research 52, 429-437. Effect of genetic variants of κ-casein and β-lactoglobulin on cheesemaking.

THE DANISH GOVERNMENT RESEARCH INSTITUTE FOR DAIRY INDUSTRY
ROSKILDEVEJ 56, DK-3400 HILLERØD, DENMARK

The Intensification - The improving purification process in sewage disposal plant

Vydrová Hana, Dairy Industry, General Management, Prague
Koubek P., Dairy Research Institute, Brno, Czechoslovakia

The effluent treatment plant in dairy factory at Bohuševice n/O. after 12 years of its operation does not correspond more to the production capacity of the dairy factory. The worse operation of the effluent treatment plant was signalized by the non-typical appearance of the sludge suspension and by the putrescent smell. This state did not correspond neither to the hygienic requirements nor to the accepted water-management standards.
To egalize the effluents and to manage the emergency peaks - the egalised tanks have been installed. The installation consists of a set of 4 covered plastic tanks with the total volume of 200 m^3. The average effluent production amounts about to 300 - 330 m^3. The plastic tanks are aerated by the blower AGKRV-GDR having capacity 665 m^3 of air per hour. The aerated plastic tanks serve like accumulating as well as egalised equipment.
The municipal effluent treatment plant in Tachov is overloaded and the demanded values of the discharged dairy effluents can not be guaranted. To resolve this situation, the installation of biofilter type Flocor was suggested to act as the pre-treatment equipment for the dairy effluent. A biofilter of dimensions: height: 5.0 m; width: 5.4 m; length: 13.8 m and volume of the charge Q = 235 m^3 has been installed on the premises of the dairy factory.
The contribution of the egalised tanks as well as of the biofilter Flocor arranged before the biological treatment plant or ahead of the municipal effluent treatment plant are evident and they are to be recommended everywhere the existing effluent treatment plant can not manage to treat the emergency peaks of hydraucally and materially inbalanced inflow of raw dairy effluents.
The both installations not only make the inflow of effluent into the effluent treatment plant more uniform and balance its pH but they reduce the quantity of the inflow impurities as well, assuring thus the functional improvement of the old effluent treatment plant. Besides, the mentioned arrangements are neither financially demanding nor time consuming.

THE APPLICATION OF THE METHOD OF DETERMINATION OF WATER TROPHIC POTENTIAL IN DAIRY WATER SYSTEM

P. Koubek, H. Vydrová+
Dairy Research Institute, Masná 5, 602 00 Brno, Czechoslovakia
+Dairy Industry, V olšinách 75, 100 98 Prague, Czechoslovakia

The determination of water trophic potential is a biological method for the evaluation of quality, degree of eutrophisation and toxicity of surface water. The principle of this method is a cultivation of green algae (Scenedesmus quadricauda) in tested water in defined conditions in special apparatus. The results are expressed in formed biomass dry matter. The tests with purified effluent and dairy drinking water were carried out. The obtained results were applied for the appreciation of possibilities applications the method for the evaluation of recipient's eutrophisation with purified effluent containing high level of nutrients (N,P).

FATTY ACIDS OF CZECHOSLOVAK SUMMER BUTTER PRODUCED IN 1968/1983

V.PALO, Slovak Technical University, Bratislava,Czechoslovakia

The content of fatty acids of Czechoslovak summer butters /June 1968,1983 respectively/ were analysed /GLC/ and compared /4,6/. The comparison were cary out especially from two territorial units of Czechoslovakia: Slovakia /SSR/ and Bohemia /CSR/.

Evaluation of the results obtained has shown that butter produced in 1983 contained a higher proportion of C_4 - C_{14} and $C_{18:2}$ fatty acids, and a lower proportion of $C_{16:0}$, $C_{18:0}$, $C_{18:1}$ and $C_{18:3}$ fatty acids than that produced in 1968. More significant differences were found in butter from Bohemian dairies than in butter from Slovak dairies.

The differences in fatty acids composition of butter produced both in the CSR and the SSR in 1983 were more signifficant than those found in 1968.

The differences obtained in fatty acid composition of Czechoslovak summer butter 1968/1983 are in agreement with the results published, for instance, in the studies dealing with Finnish butter produced in 1966/1975/1982 /1,5/, Norwegian butter from 1963/1979 /2/ or French butter produced in 1972/1982 /3/, howewer, the most significant changes in these cases concerned unsaturated fatty acids and lower ones concerned C_4 - C_{14} fatty acids. The differences are assigned to gradual genetic selection of cows, changes in their feeding, health state, etc.

It has been recommended to consider the differences found in fatty acid composition of milk fat in dairy practice, especially in butter production and its quality control.

References:

1. ANTILA V.-KANKARE V.: Milchwiss.,38,1983,478. - 2. BRATHEN G.: Meieriposten,72,1982,400. - 3. GALLACIER J.P.-KUZDZAL SAVOIE S.-BARBIER J.P.: Techn.Lait.,1984,993,13. - 4. PALO V.: Pol'nohospodárstvo,18,1972,436. - 5. RAURAMAS A.-KREULA M.-HAKARI E.-PESONEN H.: Fin.J.Dairy Sci.,34,1976,143. - - 6. REŠETKOVÁ O.: Dissertation, Bratislava 1969,SVŠT-CHTF,55p.

PRODUCTION OF UHT STERILIZED WHEY-BASED BEVERIDGES WITH HYDROLYZED LACTOSA

Ostojić,M. and Niketić,G.*
Yugoslavia,Veterinarian and dairy Institut,Novi Beograd
*Yugoslavia,A.C."Belgrade",OOUR "Standard" Padinska Skela

Whey contains ingredijents of the high nutritional value but lactose as the main solid ingredijent is not easaly digested by some people.
We hydrolyzed acid whey using the enzyme Aspergillus oryzea at 30°C and pH 4,0 - 4,2. the amount of lactose in the asid--whey which we used in ouar experiments is found to vary from 3,46 to 4,26% (TS 5,52 - 6,30%). The hydrolyzate was higher sweetness with a degree of hydrolysis of 42-66%.
Th resulting product lactose-hydrolyzed whey, we used as a component of whey-based beverages adding natural fruits. Alter UHT sterilization at 142°C the new product stay 60 days whithout changes.

THE INFLUENCE OF DIRECT AND INDIRECT UHT STERILIZATION ON THE HYDROXYMETHYLFURFURAL (HMF) CONTENT IN MILK

Niketić,G.
Yugoslavia,Agricultural Combine "Belgrade",OOUR "Standard" Padinska Skela

UHT sterilization reduces the browning reaction and the extent of browning depends largely on the process used. One of the first products of the Maillard reaction is hydroxymethylfurfural whose formation depends both on heating and storage temperature.
Whole milk (HMF content 1,75 - 3,94 nmol/l) subjected to direct or indirect UHT sterilization at 142°C for 3 s reveals higher HMF value in the case of indirect processing (HMF content 16,53 nmol/l) then for direct processing (HMF content 5,77 nmol/l).
During storage at 5°and 20°C HMF values increased and after few days slovely decrease .This increased occurs more rapidly in milk samples heted by indirect process(at 5° C is 33,57% and at 20°C is 11,81%) than in those heated directly (at 5°C is 13,34% and at 20°C is 7,45%).The greatest increased in HMF value occurs during the first few days after treatement, a period during which a rapid change has been noted in the taste of UHT sterilized milk.

PARTITIONING OF SUBSTANCES AFFECTING LIPOLYSIS IN MILK.

Introduction

Milk lipoprotein lipase (EC 3.1.1.34) is largely (80 - 90 %) associated with the casein micelles. This considerably diminishes the concentration of free enzyme and, as generally is accepted, it will reduce the turnover rate in milk.

In some samples of milk, called susceptible, cooling induces spontaneous lipolysis; in such cases more lipase is found to be associated with the fat globules.

Susceptibility of milk to lipolysis depends on the presence of a cofactor in the milk derived from the blood but also inhibitor activity may be involved.

Experimental procedure

Fresh milk from individual cows was centrifuged and the casein micelles of the skimmilk were removed. The lipase activity in the original skimmilk was tested, corrected for the influence of the separation treatment and compared to the remaining activity in the milkserum.

Test mixtures were prepared by addition of 20 % cream (20 % fat, pasteurized and homogenized) to the milk or serum; final fat content was 4 % and lipase was diluted to 83 % of the concentration in the sample.

Removal of casein micelles from skimmilk was carried out in four different ways:

I direct centrifugation at 22500 g for 90 min at 10 °C or 30 °C

II acidification to pH 4,6 with HCl and readjustment of pH with NaOH after separation at 10 °C

III rennet coagulation at 30 °C and centrifugation

IV destabilization of casein micelles by freezing in the presence of 25 mM $CaCl_2$ and centrifugation at 4 °C

Results

[illegible]

[illegible]	[illegible]	[illegible]	[illegible]
I centrifugation			
[illegible]	[illegible]	[illegible]	[illegible]
[illegible]	[illegible]	[illegible]	[illegible]
II acidification	[illegible]	[illegible]	[illegible]
III renneting			
[illegible]	[illegible]	[illegible]	[illegible]
[illegible]	[illegible]	n.d.	[illegible]
IV freezing			
[illegible]	[illegible]	[illegible]	[illegible]
[illegible]	[illegible]	[illegible]	[illegible]

n.d. not determined

Addition of a pasteurized dispersion of casein micelles to the prepared serum did not restore lipase activity.

The influence of the cofactor was tested by addition of 5 % blood serum to the skimmilk and, for control purposes, to the serum. From previous experiments binding of cofactor to casein micelles was not expected.

[illegible]

[illegible]	[illegible]	[illegible]
[illegible]	[illegible]	[illegible]
[illegible]	[illegible]	[illegible]

In an additional experiment the removed casein (by renneting) was stirred overnight in the cold with milk ultrafiltrate (MUF) with or without addition of 5 % blood serum. In the presence of blood serum a much higher part of the activity of the lipase was removed from the casein (about 350 %).

In a preliminary experiment it was found that, after addition of blood serum to milk and cooling, the cofactor was concentrated in the cream. The association between cofactor and milk fat globules was confirmed in an experiment where the milk with 2 % added blood serum was cooled, centrifuged and the skimmilk washed two or three times with fresh cream (~ 30 % fat), final fat content 4 - 6 %. Lipolytic activity of the washed skimmilk samples was estimated in freshly prepared, carefully centrifuged cream (~ 20 % fat) as a substrate; final fat content was 4 %. The decrease in lipolytic activity, compared to a control milk without added blood serum before washing is shown in the table below.

Activation of lipolysis by skimmilk with added cofactor ([illegible] % blood serum) after washing with milk fat globules. Expressed in increase of FFA in mM/100 g of fat after 24 h incubation at 4 °C

	[illegible]	[illegible]	control*
original milk	[illegible]	[illegible]	
after first wash	[illegible]	[illegible]	[illegible]
after second wash	[illegible]	[illegible]	[illegible]
after third wash	[illegible]	[illegible]	[illegible]

* [illegible]

Conclusions

The association of milk lipoprotein lipase with casein micelles has little or no influence on the lipolytic activity in milk.

Cofactor, present in blood serum and added to the milk, is able to associate with the milk lipase and to remove the enzyme from the casein micelles.

The cofactor is adsorbed to the milk fat globule membrane by cooling and this possibly causes association of part of the enzyme with the milk fat globules after cooling of susceptible milk.

Agricultural University
Department of Dairying and Food Physics
De Dreijen 12, 6703 BC Wageningen
The Netherlands

A Jellema
A Helsper
B Slaghuis

ADHERENCE OF MICROBES TO CONVENTIONAL CONTAINERS

Ashutosh Srivastava and D.N. Prasad
Ultrastructure Lab., Dairy Microbiology Division,
National Dairy Research Institute, Karnal, 132001 India

In India, Aluminium and Galvanized iron sheets are used for making cans and other milk containers. Glass bottles are still in use for distribution of milk. Attachment of spoilage organisms to these conventional containers is a major problem. The present work was conducted to study the mode of adherence of spoilage organisms to the surfaces and their total viable count was determined. Bacillus stearothermophilus was found to be the dominant organism in conventional containers used in NDRI Experimental Dairy. Standard Bacillus stearothermophilus QUM 29B obtained from Microbiology Dept., University of Queensland, Australia, was used for in vitro studies. The culture was grown in Tryptone Yeast Broth and stored in saline at 4°C. Clean Glass and Galvanized sheet pieces were dipped in sterilized skim milk containing known number of B. stearothermophilus cells and stored at ambient temperature (30°C). Viable counts were observed by Standard Plate Technique at different periods of incubation and different sanitizer levels. It was found that the viable counts increased with the increase in period of incubation and roughness of surface and decreased with increasing concentration of sanitizer. The exact mode of attachment of B. stearothermophilus (QUM) 29B to these surface conditions were studied using SEM.

A BENCH TECHNIQUE FOR EVALUATING THE CLEANING PERFORMANCE OF GENERAL-PURPOSE DAIRY DETERGENTS

D.G. DUNSMORE
HAWKESBURY AGRICULTURAL RESEARCH UNIT, NSW DEPARTMENT OF AGRICULTURE, AUSTRALIA; NOW NATIONAL DAIRY LABORATORY, RUAKURA ARC, PRIVATE BAG, HAMILTON, NEW ZEALAND.

The aim of this work was to develop a technique for assessing the cleaning performance of general-purpose dairy detergents. Glass slides were selected for the test substrates. The technique included; pre-soil substrate preparation, soiling with a milk solution and washing with test detergents. Photometric determination after pre-soil preparation, after soiling and after washing permitted calculation of Soil Removal (%) values.

An adherent milk soil was achieved by mixing calcium and magnesium chlorides with the milk solution. A double soiling with oven drying treatment was selected and developed. A washing method was developed in which the detergent was applied as a turbulent, flushing solution. Pre-detergent and post-detergent rinses were both applied. A 3% milk load was incorporated into the pre-detergent rinse and also the detergent. Detergents made for use in soft waters were tested at 300 mg/l total hardness (as $CaCO_3$), while those for hard waters were tested at 800 mg/l.

It was found that the soil removal values varied too greatly between tests to permit the use of a particular Soil Removal value as a "pass-fail" mark. This problem was solved by selecting a reference detergent with similar characteristics at the test detergents. A specific concentration of the reference was tested on every test, with the test detergents having to be statistically significantly superior to the reference to "pass".

In an experiment examining the reproducibility between two laboratories the rankings remained constant between laboratories, which permitted valid comparison of test detergents to the reference detergent within-tests and within laboratories.

The cleaning performance of a range of products offered to the dairy industry as general-purpose detergents were examined. The method was able to accurately predict the value of the detergents in the industry situation.

A BENCH TECHNIQUE FOR EVALUATING THE CLEANING PERFORMANCE OF HEAVY DUTY ALKALINE DAIRY DETERGENTS

D.G. DUNSMORE.
HAWKESBURY AGRICULTURAL RESEARCH UNIT, NSW DEPARTMENT OF AGRICULTURE, AUSTRALIA; NOW AT NATIONAL DAIRY LABORATORY, RUAKURA ARC, PRIVATE BAG, HAMILTON, NEW ZEALAND.

A bench technique had previously been developed which evaluated the cleaning performance of general-purpose dairy detergents. Glass slides were pre-cleaned and then soiled twice in a warm mixture of milk and salts, and oven-dried. The soiled slides were rinsed and then washed in turbulent flushing detergents and then rinsed again. The soil on the slides was measured photometrically after pre-cleaning, after soiling and after washing.

In this investigation a similar technique was developed for heavy-duty alkaline detergents. The same glass substrates, pre-cleaning method, washing method, photometric soil assessment and statistical treatment of the data were used. A new soiling procedure was prepared in which a milk/salts mixture was applied twice for 1h at 75°C, with a milk overflow to remove surface scum. The slides were dried at 105°C for 2h after the first soiling, and 16h after the second soiling. Detergents and rinses were prepared in water with a total hardness of 200 mg/ml (as $CaCO_3$). A reference detergent was developed for inclusion in all tests to aid in interpreation of the data.

Satisfactory intralaboratory reproducibility was achieved if, a selection was made among the soiled slides of the best slides for washing, and tests were only used if a high concentration of KOH (10 g/l) was significantly superior to a low concentration (3 g/l) within that test. The technique was satisfactorily used to determine the value of a range of detergents for use as heavy duty alkaline detergents in the dairy industry.

TUESDAY – POSTER 1

CASEIN BREAKDOWN IN CHEESE BY THE ACTION OF LIPOSOME ENTRAPPED PROTEASE

M. El Soda*, J.C. Piard, W. Alkhalaf and M. Rousseau
*Faculty of Agriculture, Alexandria University, Egypt
Laboratoire de Microbiologie Laitière, INRA, CNRZ, France

Rulactine was entrapped in 3 types of liposomes, MLV, REV and SUV which were then added to St-Paulin cheese milk. The % entrapment of the enzymes in liposomes and % retention of liposomes in cheese was in the following order MLV $<$ REV $<$ SUV. The enzyme released in the cheese hydrolysed mainly βcasein the rate of proteolysis was similar in MLV and REV and lower in SUV.

TUESDAY – POSTER 2

EFFECT OF ADDED SODIUM CHLORIDE ON TURBIDITY DEVELOPMENT IN COW'S BUFFALO'S, SHEEP'S AND GOAT'S MILKS BY ACTION OF MILK CLOTTING ENZYMES

S El-Shibiny, H F El-Dein, G A Mahran, A A Hofi, and M H Abd El-Salam
National Research Centre, and Department of Food Sci, Ain-Shams Univ Cairo, Egypt.

Sodium chloride (0-10%) was added to skim cow's, buffalo's, sheep's and goat's milks. Calf rennet, pepsin and *E.parasitica*, *M. meihei*, and M.pusillus proteases were added to milks diluted with ultra-filtrate and turbidity development was followed. Addition of NaCl decreased the initial turbidity, and turbidity development in milk treated with milk clotting enzymes. Effect of NaCl was more pronounced in cow's milk.

THE BREAK-DOWN OF PORTEINS DURING RIPENING OF CHEESE PREPARED FROM MILK TREATED WITH HYDROGEN PEROXIDE

M.Y. Ali, L.M. Buruiana*
Food Technology Dept., College of Agriculture & Forestry, Mosul, Iraq.
*Veterinary Faculty, Splaiul Independentii, 105 Bucharest, Romania.

The soluble nitrogen, the index of tyrosine and tryptophan, increased more during ripening of cheese prepared from milk treated with H_2o_2 + catalase (FH_2o_2) than cheese prepared from pasteurized milk (Fp). The β-casein broke down more in (FH_2o_2) than in (Fp), while the αs-casein was more broken down in (Fp). These modifications suggest that the H_2o_2 modifies the electron distribution on the peptide bond of casein's molecule.

Effect of soluble salts on properties of rennet coagulated milk.

R.S.Patel and H.Reuter
Institut Für Verfahrenstechnik,
der Burdestalt für Milchforschung
Kiel, West Germany.

In order to study the effect of sodium, calcium and phosphate on the properties of rennet coagulated milk, changes in rennet coagulation time (RCT), 10 min firmness (a_{10}) and time required to achieve 20 mm firmness (k_{20}) were monitored using lactodynamograph. The RCT and k_{20} increased with an increase in sodium chloride concentration in milk as well as in UF-concentrated milk, the a_{10} value on the other hand, decreased with an increase in sodium chloride concentration. RCT and k_{20} values initially decreased upon calcium addition and reached a minimum at 0.01 M calcium concentration. At high calcium concentration (0.54 M) RCT and k_{20} were severely retarded a_{10} was maximum at 0.012 M calcium chloride concentration and beyond this concentration it declined. However, with an increase in phosphate concentration RCT increased; a_{10} and k_{20} values remained almost unchanged upto the concentration of 0.65 g per litre of phosphate, but thereafter a_{10} value decreased and k_{20} value increased. Thus changes in the concentration of sodium, calcium and phosphate affected not only the rate at which coagulation occured but also the firmness of the curd.

STUDIES ON THE RIPENING OF CHEDDAR CHEESE MADE WITH Streptococcus lactis LACTOSE-NEGATIVE MUTANTS.

HYUN UK KIM, CHUNG KUG CHOI, JAI HYUN YOO*
College of Agriculture, Seoul National University, Suwon, KOREA. *College of Animal Industries, Kon Kug University, Seoul.

Lac$^-$ mutants were isolated from two strains of Streptococcus lactis and were studied for their cheese making properties, chemical and microbiological properties during ripening, for the accelerated ripening and flavour enhancement of Cheddar cheese.

Biochemical tests and sugar fermentation tests of S.lactis CH-1, S.lactis 527 and their lactose-negative (Lac$^-$) mutants revelaed no apparent differences between those of the parent strains and their mutants, except that the Lac$^-$ mutant did not ferment lactose. The phenotype of S.lactis 527, S.lactis 527-22. S.lactis CH-1, and S.lactis CH-21 were recognized as Lac$^+$ Prt$^+$, Lac$^-$ Prt$^+$, Lac$^+$ Prt$^-$, and Lac$^-$ Prt$^-$, respectively. Lac gene appeared to be located on the 34.5 and 34.0 M dal plasmid of S.lactis 527 and S.lactis CH-1, respectively.

On 15th day after manufacture, moisture, fat and protein contents of control cheese were 38, 31 and 22.3%, and moisture contents of mutants-added cheese were lower but fat and protein contents of mutants-added cheese were higher than those values of control cheese. No dramatic pH or acidity changes of cheeses were found during ripening period, which varied between pH 4.9 and pH 5.1.

Water-soluble nitrogen contents of cheeses were increased with the ripening time and no big differences were found between two experimental cheese groups. PTA-soluble nitrogen contents of cheeses were increased with the ripening, PTA-soluble nitrogen contents of mutants-added cheese were 1.5-2.4 times higher than that of the control cheese. Free amino acid contents of control and mutants-added cheese after three months of ripening were 47.56 μmole/g and the range of 57.93-84.52 μmole/g.

There were no big differences in texture of all cheeses during three months of ripening. When the 3 month-ripened cheeses were evaluated by the panel, mutants-added cheeses showed higher acceptibility in flavour and texture but there was no significant difference among mutants-added cheeses. In conclusion, mutants-added cheeses showed higher level of free amono acid contents and higher acceptibility in flavour than the control cheese, which tells the favorable acceleration of cheese ripening.

ISOLATION AND SOME PROPERTIES OF LOW-TEMPERATURE-ACTIVE PROTEINASE FROM COMMERCIAL CHEDDAR CHEESE

R.Akuzawa and K.Yokoyama
Depertment of Animal Products Science and Technology,
Nippon Veterinary and Zootechnical College,
Musashino-shi, Tokyo 180, Japan

A Proteinase which was showed maximum activity at low temperature was purified from the phosphate buffered homogenates of New Zealand cheddar cheese.
The results obtained through the purification process are summarized in table. The purity of purified enzyme was detected by the disc-electrophoresis. The enzyme showed a single band and the enzyme activity was detected at the same position. The molecular weight of enzyme was estimated to be 12000. Its maximum activity was found at pH 6.5 and 10°C. We previously isolated a proteinase from lactic acid bacteria, which showed a maximum proteolytic activity at a low temperature level(5~10°C) and in a slightly acidic or neutral pH range(1,2). The low-temperature-active proteinase(LTAP) isolated from the cheese in the present experiment had properties similar to the LTAP isolated previously from lactic acid bacteria. We concider that the LTAP isolated from the cheese originated from the starter organism. Detailed studies seeme necessary on the properties of lactic acid bacteria isolated from cheese.

Table 1. Purification of proteinase

Step of purification	Total* protein (mg)	Total* activity (units)	Specific activity (units/mg)	Purification	Yield of activity (%)
Cheese-homogenate extract	8.02	102.38	12.77	1	100
Sephadex G-200 chromatography	4.08	83.31	20.42	1.6	81.4
DEAE-Sephacel chromatography	0.52	46.55	89.52	7.0	45.5
Sephadex G-50 chromatography	0.41	41.49	101.20	7.9	40.5
Sephadex G-50 chromatography	0.38	39.30	103.44	8.1	38.4
Disc-PAGE gel-slice extract	0.01	5.1	510.11	39.9	5.0

* obtained from 1 g of cheese.

(1) Akuzawa,R. and Yokoyama,K.,(1982) Jpn.J.Zootech.Sci., 53,814. (2) Akuzawa et.al.,(1983) ibid., 54,685.

EFFECT OF FORMALDEHYDE AND LYSOZYME ON LACTIC ACID BACTERIA DURING GRANA CHEESE RIPENING

R. LODI, R.TODESCO, A.M.VEZZONI, M.LANZANOVA

Centro Latte C.N.R.-*Industrie Agrarie - Università di Milano - Via Celoria, 2
20133 MILANO ITALY

The use of formaldehyde, at approximately 25 ppm, has been allowed by Italian law from decades to prevent late blowing in Grana Padano and Provolone cheeses. As the first results showed that lysozyme was adequately effective to prevent butyric fermentation in cheese, since 1983 the use of lysozyme has been allowed for a trial period of three years by Italian law in Grana Padano, Provolone, Montasio and Asiago cheeses. Amounts of lysozyme in cheese milk must not exceed 25 mg/l, provided that there is no residue of lysozyme greater than 300 ppm in the cheese.

Due to its known indiscriminate, bacteriostatical action, formaldehyde also effects lactic microflora. On the other hand, several AA. verified that *in vitro* lysozyme does not influence lactic acid bacteria, with the exception of some *L. helveticus* strains. At the moment, there is no evidence that lysozyme acts on lactic microflora evolution during the Grana cheese ripening.

Materials and Methods

In a farm of the Padana plain two Grana cheesemaking processes were carried out using the same milk, to which formaldehyde had been added during the rest in the vessel. At the beginning of the process lysozyme was added to only one of the two vats (25 ppm), while the technology and the starter were exactly the same for both of them.

At the same time a third cheesemaking was carried out: formaldehyde was not added to the milk in the vessel and lysozyme was not added to the milk in the vat.

The following bacteriological analyses were carried out on milk, curd and on 1 and 3 month ripened cheese:

- lactic acid bacteria count made in MRS agar (Difco). Incubation: 37°C for 2 days;

- lactic acid bacteria isolation made in MRS broth (Difco):37°C for 2 days. After purification of all the grown colonies, the lactic acid bacteria were classified with API system and according to Sharpe (1979).

Results

Results in Fig.1-a show that formaldehyde greatly reduces lactic acid bacteria during the rest in the vessel: from 40×10^6 to approximately 10×10^4 in added milk.

The curds also show the same remarkable difference (Fig.1-b): 40×10^6 in the one not added and approximately 20×10^4 in both curds with formaldehyde (with and without lysozyme).

After 1 and 3 months of ripening (Fig.3-c and -d) the microbical counts can be considered equal: nearly $10 - 15 \times 10^7$ after 1 month (at the end of the brine salting) and $10 - 15 \times 10^6$ after 3 months of ripening.

Concerning the distribution of the various microbical species, it is possible to notice that L. bulgaricus, L. lactis, L. fermentum, Str.thermophilus Str. lactis are present in all of the curds.

These species are present in an approximately equal numerical ratio in the two curds with formaldehyde (with and without lysozyme).

In the one without additives there is a greater prevalence of Str. thermophilus.

At the end of the brine salting the lactic microflora is reduced; only a little number of species is present: L.casei-rhamnosus, L.casei, L.plantarum and Pediococci. In the cheese without additives L. fermentum can also be found.

After 3 months of ripening L. casei-rhamnosus, L. casei, Str. thermophilus and Pediococci are still present in all of the three cheeses. L. fermentum is still present in the cheese without additives. These species are nearly always present in the same numerical ratio.

Conclusions

The results confirm the following assertions:

- formaldehyde greatly inhibits microflora, including lactic microflora; this inhibition can considerably reduce the more useful lactic acid bacteria, in the early stages of the Grana cheese ripening;
- lysozyme does not change the composition and the activity of the lactic acid bacteria, therefore allowing a correct ripening.

PRODUCTION OF SOMBORSKI CHEESE BY ULTRAFILTRATION METHOD

M. CARIC, D. GAVARIC, S. MILANOVIC, LJ KULIC
Faculty of Technology, V.Vlahovica 2,
2100 NOVI SAD, Yugoslavia

Somborski cheese is one original product among many Yugoslav indigenous types of cheese. The main features (1) in production of the cheese type are ; high temperature of raw milk pasteurization (90-95°) with the aim of whey protein denaturization, addition of 20% of water to cheese milk and soaking the pieces of cheese curd in water to reduce lactose content, packing in wooden buckets, etc. The product obtained has soft body, creamy consistency, pleasant, mild flavour and taste.

The growing interest for application of ultrafiltration (UF) in cheese making and a lot of laborious, manual work during classical production of Somborski cheese were the main reasons to conduct investigations for UF application (MMV method (2) in the production of this cheese.
The main differences between the traditional method of Somborski cheese production and the new one, were omission of high pasteurization (serum proteins are already included in retentate), and usage of diafiltration (to diminish lactose content in cheese curd).

Main operations in Somborski cheese production are : raw milk → standardization → pasteurization → cooling → ultrafiltration → diafiltration → cooling → renneting → cutting(forming) → water soaking → dry salting → packaging → pressing → ripening → Somborski cheese.

Table 1 : PHYSICO CHEMICAL COMPOSITION OF CONTROLS AND UF SOMBORSKI CHEESES AT THE END OF RIPENING

COMPONENTS	CONTROL	TYPE OF CHEESE SAMPLE A_1	A_2	B_1	B_2
Dry matter	49.95	49.90	50.81	49.40	48.35
Milk fat	27.00	26.00	28.50	30.50	27.00
Total proteins	19.76	19.45	18.48	15.69	18.05
Lactose	0.10	0.18	0.15	0.14	0.10
Ash	2.96	4.33	3.87	3.10	3.38
Calcium, mg %	407.00	445.00	436.00	388.00	370.00
Phosphorus, mg %	172.00	151.00	158.00	132.00	134.00
Ripening index Shilovich	55.00	60.00	45.00	40.00	45.00
Titrable acidity °SH	76.00	73.00	75.00	53.00	74.00
pH	4.95	4.95	4.95	5.00	4.95

Note: Sample A_1 - without diafiltration; without water soaking
Sample A_2 - without diafiltration; with water soaking
Sample B_1 - with diafiltration; without water soaking
Sample B_2 - with diafiltration; with water soaking.

At the end of the ripening period dry matter content UF investigated cheeses were similar to control especially samples A_1 and B_1. Variation of fat content in cheeses was due to batchwise production, as well as salt content. There was negligible difference in lactose content, titrable acidity and pH among cheeses, due to small drainage of whey during pressing of acid curd, calcium and phosphorus content in UF cheeses was similar to control.

Table 2 : ABSOLUTE AND RELATIVE PARTS OF INDIVIDUAL PROTEIN FRACTIONS IN CLASSICAL AND UF CHEESE AT THE END OF RIPINING

NITROGEN FRACTIONS %	TYPE OF CHEESE SAMPLE				
	CONTROL	A_1	A_2	B_1	B_2
ABSOLUTE PART					
TN	3.10	3.05	2.88	2.46	2.83
NPN	0.215	0.234	0.244	0.181	0.252
NCN	0.396	0.483	0.621	0.532	0.714
TN x 6.38	19.76	19.46	18.48	15.69	18.05
"True proteins"	18.41	17.97	16.82	14.54	16.45
Casein	17.25	16.38	14.46	12.30	13.50
Serum proteins + albumose+peptone	1.16	1.59	2.41	2.24	2.95
RELATIVE PART					
TN	100.00	100.00	100.00	100.00	100.00
NPN	6.94	7.64	8.47	7.36	8.90
NCN	12.77	15.84	21.56	21.62	25.23
"True proteins"	93.07	92.34	91.56	92.67	91.13
Casein	87.21	84.17	78.44	78.39	74.79
Serum proteins + albumose & peptone	5.86	8.17	13.11	14.40	16.34

Special point of interest is mutual relationship among different nitrogen fractions in cheese produced. UF cheeses obviously posess significantly more serum proteins than control (sample B_2 almost triple), which gives much higher nutritive value to these samples.

Organoleptic evaluation revealed good quality of all UF cheeses, among them sample B_2 resembled the most the original cheese. Investigations are continuing for further modifications in cheese production by ultrafiltration and establishing exact economic advantages.

Literature :
1. Caric M., Milanovic S., Garavic D., Litvai V., (1983): Zbornik Radova, TechnoloskiFakultet, Novi Sad, 14.
2. Maubois J L., Mocquot G.,Vassal(1971) : Le Lait, 495

INFLUENCE OF USE OF GLYCERIDE COATING MATERIALS ON PROTEOLYSIS IN TRAPIST CHEESE

M. CARIC, L.J. KULIC, D. GAVARIC, S. MILANOVIC, Z. VRBASKI, K. VRANAC
Faculty of Technology, Faculty of Science
V. Vlahovia 2, 21000 Novi Sad, Yugoslavia

In modern technology different ways of cheese surface protection from mold growth, during cheese ripening, are used. Edible acetomono-glyceride cheese coating of domestic (Yugoslav) components was produced in laboratory conditions and its efficiency in cheese protection is investigated in industrial Trapist manufacture. Examination of the influence of coating material on chemical composition and yield during 50 days ripening period showed excellent yield:0.94% weight loss; compared to 12% weight loss by plastic coating, or 12.6% without coating (1).

TABLE 1: The changes of total proteins (TP), soluble proteins (SP) and non protein nitrogen (NPN_2) in dry matter (%) of cheeses, coated with plastic during ripening period.

Ripening Period (Days)	Plastic coating material: TP (%) 1	TP (%) 11	SP (%) 1	SP (%) 11	NPN_2 (%) 1	NPN_2 (%) 11
10	43.35	43.97	2.54	2.17	1.47	1.59
20	43.16	43.07	4.44	3.74	1.67	1.71
30	42.85	42.68	4.56	4.77	2.40	1.84
40	42.44	42.30	5.15	5.11	2.48	2.07
50	42.18	42.05	6.28	6.18	3.86	3.70

1 - Plastic coating material Mowilith (Yugoslavia)
11 - Plastic coating material AGD (Holland)

The share of total proteins in dry matter decreased slightly during 50 days of Trapist ripening under plastic coating (Table 1). Protein degradation resulted in soluble proteins, as well as non protein nitrogen increase. Soluble proteins increase was at its greatest after 20 days of ripening being about the same for both plastic coatings.

By control sample (without any coating) total proteins in dry matter amounted 45.25% and soluble proteins 1.85% after production and 42.59% and 3.85%, respectively, after 50 days of ripening.

TABLE 2: The changes of total proteins (TP), soluble proteins (SP) and non protein nitrogen (NPN_2) in dry matter (%) of cheeses, coated with acetylated glyceride during ripening period

Ripening Period (Days)	Acetylated glyceride coating material TP (%)		SP (%)		NPN_2 (%)	
	1	11	1	11	1	11
10	44.53	44.32	1.67	2.30	0.74	1.51
20	44.25	43.82	4.05	5.38	1.20	1.70
30	43.63	43.55	5.34	5.62	2.26	2.45
40	43.03	42.17	5.40	5.99	2.60	2.72
50	42.47	39.62	7.57	5.91	5.15	5.41

1 and 11 - Acetylated glyceride coating material, experimental samples : 1 and 11.

Comparison of the results for total proteins, soluble proteins and non protein nitrogen changes, in cheese coated with acetylated glyceride during ripening (Table 2), with results given in Table 1, shows less expressed protein degradation by plastic coatings. This is also confirmed by electrophorectic pattern of proteins in different coated Trapist cheeses.

Polyacrylamide (PAG) electrophoretic examination in following patterns of casein fractions :

Cheese without coating
$\alpha_s : \beta$: 111 : 11 : 1 = 47.74 : 46.40 : 3.92 : 1.50 ;:0.43

Acetylated glyceride coating material
1 -$\alpha_s : \beta$: 111 : 11: 1 = 62.40 : 34.42 : 1.89 : 1.06 : 0.23
11-$\alpha_s : \beta$: 111 : 11: 1 = 58.11 : 36.17 : 3.28 : 2.21 : 0.04

Plastic coating material
1 -$\alpha_s : \beta$: x = 62.54 : 33.89 : 3.56
11-$\alpha_s : \beta$ = 62.29 : 37.71

Acetylated glyceride coatings resulted in higher casein proteolysis degree in cheese, similar to natural process (cheese without coating), compared to plastic coatings.

Investigation of the effect of acetomonoglyceride cheese coating on protein degradation during ripening showed no big differences compared to non-coated Trapist, while protein changes were less intensive in plastic coated cheeses. Cheeses with experimental coating underwent proteolysis more rapidly and thoroughly than cheeses coated with plastic coatings, which is confirmed with TP, SP and NPN_2 contents, as well as with the electrophoretic pattern.

REFERENCE:

1. CARIC M., Vrbaski Z., Kurlic LJ., Gavaric D., Vranac R., Dovznoev Z. (1985) : Mljekarstvo 9, 267.

INVESTIGATIONS OF POSSIBILITIES OF RAISING THE SHARE OF WHITE PICKLED CHEESE FOR PROCESSED CHEESE PRODUCTION

M. CARIC, D.GAVARIC, S.MILANOVIC, L.J.KULIC
Faculty of Technology, V.Vlahovica 2, 21000 Novi Sad
Yugoslavia

The initial idea of cheese processing in order to utilize cheeses which would not be suitable for direct consumption is now applied to a wide variety of processed cheese types witn increasing total world production. The aim of these experiments was to decrease production costs in processed cheese manufacture by inclusion of white brined cheese in the blend, up to the highest quantity which has no adverse effect on final product quality.

Table 1. INGREDIENTS USED IN THE MANUFACTURE OF PROCESS CHEESE

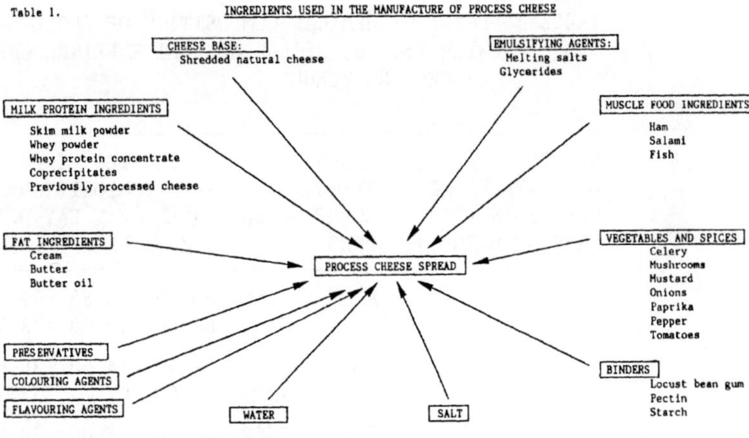

Processed cheese samples were industrially produced by usage of 2 different commercial emulsifying agent mixtures.
The operations in processed cheese production are:
Selection of natural cheese ➝ computation of ingredients ➝ blending ➝ shredding ➝ addition of emulsifying agent ➝ processing ➝ (homogenization) ➝ packaging ➝ cooling ➝ storage.

Designing the proper blend composition is of high priority in processed cheese manufacture. Blending is strongly influenced by the properties of the product wanted (blocks, spreads, slices, foods, processed cheese analogues). Cheese used in blends are usually of different type, age and consistency.

TABLE 2. AVERAGE COMPOSITION OF WHITE BRINED CHEESE

TYPE OF WHITE BRINED CHEESE	FAT IN DM (%)	DM (%)	Total PROTEIN	NaCl (%)	PH OR ACIDITY(°)
FETA(Greece)	48.52	46.37	-	5.05	4.3-4.4
BJALO SALAMURENO	27-31	40-42	10-12	3-4	290-320
SIRENE(Bulgaria)	43-45	42-44	16-18	2-3	280-300
BELI SIR u KRISKAHA	45-50	52-56	-	3-5	-
(Yugoslavia)	47.78	45.56	19.19	2.22	252
BRINZA(USSR)	45-50	42.00	-	4-10	-

Having high NACl level, as it is evident from table 2, white brined cheese is usually used only up to 20% in process cheese spreads, negative effect of NaCl is especially exposed in flavour and consistency of the final product.

TABLE 3. PHYSICO-CHEMICAL COMPOSITION OF PROCESSED CHEESE WITH 25, 30, 35 AND 40% WHITE BRINED CHEESE IN PROCESSING BLEND

RELATIVE SHARE OF WHITE BRINED CHEESE IN PROCESSING BLEND,%	EMULSIFYING AGENT USED	PHYSICO -CHEMICAL COMPOSITION				
		pH	WATER (%)	FAT (%)	FAT IN DM (%)	NaCl (%)
25	KSS[a]	5.66	60.25	15.65	39.37	1.35
	SOLVA[b]	5.65	61.50	14.90	38.70	1.38
30	KSS	5.67	62.40	14.50	38.60	1.33
	SOLVA	5.78	62.60	13.80	37.10	1.24
35	KSS	5.62	61.80	15.00	39.26	1.36
	SOLVA	5.65	62.10	14.50	38.25	1.35
40	KSS	5.59	62.90	14.90	39.10	1.69
	SOLVA	5.50	63.00	14.00	37.80	1.75

[a] Emulsifying agent mixture (KSS-1 : KSS-2 = 2 : 1) Produced by Koteksprodukt, Novi Sad

[b] Emulsifying agent mixture (Solva 820 : Solva 740-2:1) Produced by Giulini Chemie, Ludwigshafen

Quality of processed cheese spreads with different amounts of white brined cheese in the blend (20-40%) is shown in table 3. Sodium ion concentration in white brined cheese has been decreased by diffusion, prior to blending.

Both Yugoslav (KSS-1 and KSS-2) and imported commercial emulsifiers showed excellent emulsifying ability, also is the presence of high percentage (40%) of white brined cheese.

Results of organoleptic evaluation showed that even as high quantity as 40% of white brined cheese could successfully be used in processed cheese production, provided that diffusion prior to blending takes place.

References: 1. Caric M., Gantar M., Kalab M. (1985) : Food Microstructure 4.297.

2. Caric M., Milanovic S., Gavaric D., Kulic I.J., (1986): Mlekarstvo 2,49

PURIFICATION AND CARACTERIZATION OF A CELL-WALL PROTEASE FROM STREPTOCOCCUS LACTIS

V. MONNET, D. LE BARS, J.C. GRIPON
Laboratoire de Biochimie et Technologie Laitières
Institut National de la Recherche Agronomique
78350 JOUY EN JOSAS FRANCE

Mesophilic streptococci have complex nitrogen requirements and the concentration of free amino-acids and peptides in milk is too low to support an optimal growth. Cell-wall proteinases are essential for casein hydrolysis and peptides production. The present communication describes the purification and characterization of a cell-wall proteinase from *Streptococcus lactis* .

A proteolytic activity, detected with a radioactive substrate (^{14}C methylated casein or hemoglobin) is spontaneously released when the cells, after growth in milk, are incubated in a buffer without calcium. Presence of peptides and/or absence of calcium in the growth medium seem to repress this proteolytic activity production.

The released activity seems to be the same for the five strains of *S.lactis* we tested. This activity has been purified for the strain *S. lactis* NCDO 763. Only one proteinase, the purity of which has been shown by electrophoresis, has been detected after 3 steps of chromatography : ion exchange, molecular sieving and chromatofocusing.

This cell-wall proteinase has original properties. It is a serine proteinase, active at acid pH (6-6,5 on casein and 4,8 on hemoglobin) and with a high molecular weight (80.000). Its optimal activity as been determined at 35°C and it is more active on β-casein than on the other caseins. Five peptides liberated from β-casein, have been identified ; they are located in the C-terminal part of the molecule and are peptides 167-175, 176-182, 183-193, 194-207 and 194-209. All of them can be bitter and other authors have observed a bitter flavor with a peptide located in position 194-209 in the sequence of β-casein. Owing to their size, these peptides have probably to be hydrolyzed by peptidases before being transported across the cytoplasmic membrane.

PRODUCTION METHODS FOR DANBO, HAVARTI AND DANABLU CHEESE MADE FROM UF-CONCENTRATED MILK.

K. B. QVIST, D. THOMSEN,
K. FORSINGDAL and G. HYLDIG

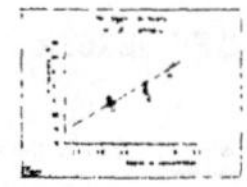

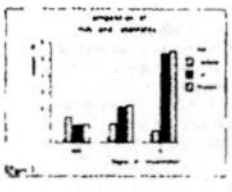

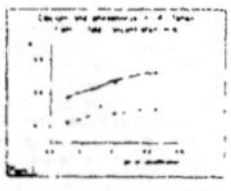

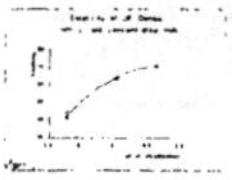

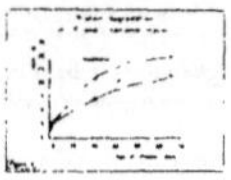

ABSTRACT

UF-production methods for the traditional Danish cheese types Danbo, Havarti and Danablu made from 2-5 fold concentrated milk are discussed and so are the key variables used to control moisture, acidity and mineral content. Also, the dependence of yield on the degree of concentration is discussed. Quality of all varieties has been from satisfactory to very satisfactory.

INTRODUCTION

In Denmark the most produced cheese variety, Feta, is made from UF-concentrated milk. The reason for the popularity of the UF-method is that it offers several advantages over the traditional method: more efficient utilization of milk protein (fig. 1), less rennet is used and the process is much more labour efficient.

However, UF-cheesemaking also presents problems. Due to the composition being very different from that of milk (fig. 2), the UF-concentrated milk has cheesemaking properties very different from those of natural milk (Culioli and Sherman, 1978). Major modifications of the production methods are necessary, especially as the degree of concentration rises. We have directed our attention towards developing UF methods for traditional Danish cheese types such as Danbo, Havarti and Danablu. The following sections describe some of our results in this area.

DANBO CHEESE FROM 2 FOLD CONCENTRATED MILK

Preferred process: LTST pasteurized whole milk is pH regulated to 6.40, then concentrated 2 fold and diafiltrated to a lactose content of ca. 3.4%. The retentate is LTST pasteurized.

The cheeses are made in vats using a comparatively traditional technique. Salt (0.2%), starter (ca. 1%), salpetre (0.015%) and standard rennet (0.025%) are added to the retentate and coagulation takes place at 30 C. Addition of water for scalding is 40% and the scalding temperature is 38 C. Pressing, salting and storage are performed as in the traditional process.

pH at ultrafiltration, amount of water for diafiltration, amount of starter, scalding temperature and amount of water added in the vat are the variables used to control moisture, acidity, mineral content (fig. 3) and body (fig. 4) of the cheese.

Quality: Body and flavour are fully satisfactory after 6 weeks of storage. However, after 16 weeks of storage some of the cheeses are given remarks such as mealy body and off-taste.

Yield: Recovery rate for fat is rather low, 85-87%, which should be seen in relation to usage of inadequate equipment for cutting and stirring. The recovery rate for nitrogen is about 1 higher than for the traditional method.

HAVARTI CHEESE FROM 5 FOLD CONCENTRATED MILK

Preferred proces: LTST-pasteurized milk is concentrated 5 fold using UF and diafiltrated to a lactose content of ca. 2.2%. The retentate is LTST-pasteurized and DVS-starter (ca. 0.1%), salpetre (0.015%) and standard rennet (0.05%) are added. Coagulation takes place in an Alcurd coagulator (Alfa-Laval) at 38 C. At cutting the cheese grains are dropped directly into Havarti moulds The moulded cheeses are kept in trolley tables at 30 C for 2 hours and turned periodically. Cooling, salting and storage are performed as in the traditional process.

pH of milk at ultrafiltration, amount of water for diafiltration, amount of starter and temperature in trolley tables are the principal variables used to control moisture, acidity and mineral content of the cheeses.

Quality: With respect to rind, colour, texture and body the UF-Havarti is of very good quality. Flavour and taste are somewhat milder than typical of the traditional product, undoubtedly due to a slower degradation of protein (fig. 5) in accordance with the results of de Koning et al. (1981).

Yield: Recovery rate for nitrogen is 84% as opposed to 76% (Birkkjær and Thomsen, 1984) in the traditional process. The consumption of skim-milk is about 9% less with the UF-process

DANBO CHEESE FROM 5 FOLD CONCENTRATED MILK

Preferred process: Production of retentate, addition of DVS-culture, salpetre and rennet are performed as with the Havarti process. Coagulation takes place in a modified Alcurd coagulator at 32 C. At cutting the cheese grains enter a vat containing UF-permeate. The curd is stirred for 15-30 min. and scalded to about 40 C Pressing, cooling, salting and storage are performed as in the traditional process.

pH of milk at ultrafiltration, amount of water for diafiltration, amount of starter, scalding temperature and stirring time are the principal variables used to control moisture, acidity and mineral content of the cheeses.

Quality: Using the technique indicated Danbo cheese of a high quality has been made. However, UF-Danbo like UF-Havarti has a somewhat milder taste than the traditional product.

Yield: Yield is less than in the UF-Havarti process due to loss of whey protein during stirring in permeate. It should, however, be possible to collect the whey from the process and re-use the protein. Alternatively, it may be possible to eliminate the need for stirring in added liquid and thus obtain the same yield as in the UF-Havarti process.

DANABLU CHEESE FROM 3-5 FOLD CONCENTRATED MILK

Preferred process: Skimmilk and homogenized cream are standardized to 4% fat. This cheese-milk is thermizated to 62 C and concentrated 3-5 fold using UF. The retentate is pasteurized to 62-72 C and DVS-starter (ca. 0.1%), mould and standard rennet (0.025%) are added. Coagulation takes place in an Alcurd Coagulator at 35 C. After cutting the cheese grains are mixed gently for 2-5 min. before the curd is put into moulds. During the draining period (18 hours at 25 C) the cheeses are turned 3-4 times. Salting, pinching and storage are performed as in the traditional process.

pH at ultrafiltration, degree of concentration, amount of starter, temperature at drainage and during brine salting and concentration of brine are the principal variables used to control moisture, acidity and mineral content.

Quailty: With respect to colour, texture, body and taste the UF-Danablu is of good quality. The rind tends to be somewhat open and the cheese grains do not fuse as well as they do in the traditional product.

Yield: The recovery rate for nitrogen depends on the degree of concentration and varies from 80-84% as the concentration varies from 3.5 to 5 (fig. 1). The recovery rate for fat is 99%. This gives rise to a considerable inprovement in fat economy as it is not possible to recover the whey fat in the traditional process due to homogenization of the fat.

REFERENCES

Birkkjær, H. E. and Thomsen, D.: Mælkeritidende 97 (1984) 84-85.

Culioli, J. and Sherman, P.: Journal of Texture Studies 9 (1978) 257-281.

de Koning, P. J., de Boer, R., Both, P. and Nooy, P. F. C.: Netherlands Milk and Dairy Journal 35 (1981) 35-46.

THE DANISH GOVERNMENT RESEARCH INSTITUTE FOR DAIRY INDUSTRY
ROSKILDEVEJ 56, DK-3400 HILLERØD, DENMARK.

GROWTH AND ACID PRODUCTION OF LACTIC STREPTOCOCCI IN ULTRAFILTERED MILK

HERLEV JENSEN

Abstract

Growth and acid production of mesophilic lactic streptococci were studied in ultrafiltered milk retentates concentrated 2 1 and 5 1. The buffer capacity of such retentates is quantified As well growht as acid produced increased with the concentration of dry matter in the retentate. Furthermore S. lactis 10 045 produced lactic acid to a higher level in retentate than in milk. Magnesium activated the acid production with 5-10 % at 30°C and 2-5 % at 37°C

Introduction

Manufacture of semi-hard cheese from milk concentrated by ultrafiltration has now been developed to such a level that this method will be preferred in the very near future Until now there have been a number of reports on the manufacture of semi-hard cheeses from milk concentrated by ultrafiltration Covacevich et al (1978), Maubois et al (1975) In this connection there has been some difficulties with development of a sufficient activity of the starters used because the starter has other growth conditions in the concentrated medium but also because a temperature at 35-37°C is wanted. This temperature is higher than temperature optima for the mesophilic streptococci In order to obtain a little more information about the growth and acid production in retentates some investigations have been carried out in that question

Material and methods

Cultures

In this investigation were used a strain of S lactis 10.045. The strain was maintained by daily subculture in sterile skim milk autoclaved at 121°C for 15 min As incubation temperature was used 30°C for 16-18 hours

Milk-retentates

Milk retentates were prepared from whole milk standardized to 3.4 % fat. The milk was pasteurized at 72°C for 15 sec , cooled and ultrafiltered at 52°C using a DDS M 36-9-3-6A 61 PP plant to a concentration of 24 % and 40 % total solid equal to a concentration factor 2 1 and 5 1. The retentates were then pasteurized at 65°C for 15 sec , cooled to 30-35°C and used directly or added permeate

Growth studies

Growth and acid production were investigated for S. lactis 10 045 in milk, retentates and retentates added permeat to an amount equal to the milk before ultrafiltration As inoculum was used 4 % and as incubation temperature 30°C or 37°C for 6 hours after which the number of bacteria (cfu) and the amount of lactic acid were determined Furthermore the growth and acid production were measured at a steady state pH 6 0, at 30°C with an inoculum of 4 % culture

Buffer capacity

The buffer capacity of the milk and milk retentates was determined by the method of Sutherland et al (1981)

Magnesium as activator for the acid production

As activator for the acid production 50 ppm of magnesium was used

Analysis of milk, milk retentate and permeate

The composition of milk, milk retentate and permeate was analyzed and the results are shown in table 1 as a typical composition of the media used It can be seen that the composition of the milk and the milk made of retentate + permeate is almost the same

Results

Growth and acid production

The effect of ultrafiltration on the growth and acid production of S lactis 10 045 has been carried out in milk and retentate added permeate in such an amount that the medium was equal to the milk before ultrafiltration

The results showed that growth as well as acid produced were highest in the medium made of retentate + permeate

The investigations concerning growth and acid produced in retentate and retentate + permeate under controlled pH-condition showed that the number of cells was 2 times higher in retentate than in retentate + permeate. The lactic acid produced was about 210 mM in retentate and 75 mM in retentate + permeate In retentate + permeate the lactic acid increased proportionally more than the number of cells The maximum specific growth rate was calculated to be respectively 0,73 h^{-1} and 0.68 h^{-1}

Buffer capacity

The buffer capacity in retentate 5 1 and 2 1 was 104 and 82 µmol lactic acid per pH unit per g milk medium In retentate + permeate and milk the buffer capacity was 40 and 47 µmol lactic acid per pH unit per g milk medium

Influence of magnesium on the acid production

The investigation is carried out at 30°C and 37°C with and without adding magnesium to the retentate 5 1

Addition of 50 ppm magnesium per ml retentate increased the acid production of S. lactis 10 045 at 30°C about 5-10 % and at 37°C about 2-5 %.

References

Covacevich, H. R and Kosikowski, F V (1978) J. Dairy Sci 61, 701

Maubois, J L and Mocquot, G (1975) J Dairy Sci. 58, 1001

Sutherland, B J and Jameson, G. W (1981) Aust J Dairy Technol 36, 136

Chemical composition of milk retentate and permeate

%	Milk	Retentate + permeate	Retentate 5 1	Retentate 2 1	Permeate
Dry matter	12,0	12 0	35-36	18-19	5 3
Lactose	4,8	4 45	2,1-2,2	4,2	4,45
Protein	3 5	3 5	15-16	6-7	0 25
Fat	3 1	2,9	14-15	6-7	0
Citrate	0 18	0 18	0 18-0,17	0 21-0.22	1200
Ash	0,70	-	1,6	-	-
Calcium	0 12	0 13	0 50	0 21	0,30
Fosfor	0 10	0,13	0 30	0,15	0 40

Buffercapacity in milk and retentate

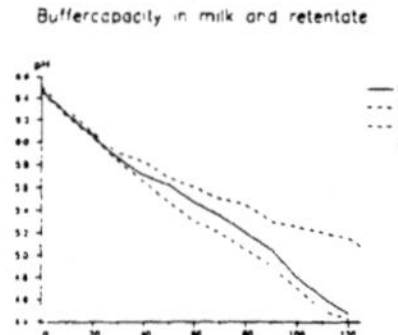

Growth and acid production of S.lactis. 30°C, 6 h

Medium	CFUx10⁶/ml	Lactic acid mM
Milk	110	48
Recomb milk	160	54

Growth of S lactis, pH 6 0

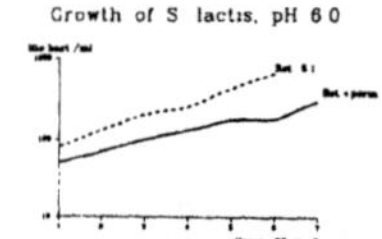

The effect of 50 ppm magnesium on the acid production in percent

Culture	30°C	37°C
Starter 1	33	2
- 2	11	7
- 3	4	7
- 4	6	0
S.arom. 1	3	2
- 2	3	4
- 3	6	6
- 4	19	0

Retentate 5 1
Inoculum 3 pct – 5 h

THE DANISH GOVERNMENT RESEARCH INSTITUTE FOR DAIRY INDUSTRY
ROSKILDEVEJ 56, DK-3400 HILLERØD, DENMARK.

ACCELERATING THE RIPENING AND IMPROVING THE FLAVOR OF EMMENTHAL CHEESE WITH SELECTED LACTOBACILLUS CASEI-CULTURES

Säde Mantere-Alhonen, Hanna Mäkäräinen and Matti Antila
Department of Dairy Science, University of Helsinki
Viikki
SF-0071 Helsinki
Finland

ABSTRACT. The aim was to isolate and combine new active *Lactobacillus casei*-cultures for emmenthal cheese. Acid production, proteolytic activity and physiological properties of the strains were of main interest. The new starters were used in normal cheese manufacture. The flavor of the test cheeses was clearly better than that of control cheeses. By combining new active starters we have succeeded to improve the flavour and to accelerate the ripening of emmenthal cheese.

The ripening of emmenthal cheese is known to be long. It is produced of non-pasteurized milk with an authentic microflora with its own problematic, but there still exists possibilities of developing new startercultures. In this work special attention has been paid to the lactobacilli because of their high amount and rich assortment in ripening cheese. Even though it is well known, in different countries bacterial startercultures and especially *Lactobacillus*-species are throughly investigated and developed. However, rather few publications occur (some of them to be mentioned: Langsrud & Reinbold 1973, Steffen 1982, Accolas & Auclair 1983, Turner 1983 and Thomas 1985). Therefore it was our aim to isolate and combine new active lactobacilli cultures for emmenthal cheese and especially *Lactobacillus casei*-cultures which have remained with less attention in cheese making.

The acid production and proteolytic activity of lactobacilli were of main interest in this investigation. The organisms in question were selected from cheese material giving a representative picture of lactobacilli. 35 different strains were isolated using MRS-agar and identified according to cell shape and physiological properties: growth at 15° and $45^{o}C$, carbohydrate fermentation, reactions in litmus milk, NH_3 formation from arginine, and proteolysis. Further 17 strains were chosen for testing their acid production in skim milk at $40^{o}C$ during 10 hours. In the proteolytic activity of these strains there were no big differences, it was strong or very strong. The main part of lactobacilli isolated were *Lactobacillus casei*-strains followed by *Lb. lactis* and *Lb. helveticus*.

The Lb. casei-strains were slower acid producers than the two others. The aroma production was simultaneously controlled. Differences in the properties of the isolated Lb. casei-strains occured in the cell shape, the growth at 45°C and the ability to ferment rhamnose, esculine and cellobiose.

Basing on the properties investigated several selected Lb. casei-strains were combined to new active startercultures to be used in normal Finnish emmenthal cheese manufacture. In the time programm of cheese making variations caused by the development of acidity have not been taken into consideration, but the process was just the same as normally in Finland. In addition to the usual cheese evaluation after three months of ripening a separate aroma test was carried out and the taste of the test cheeses was estimated to be clearly better than that of the control cheese, and the ripening time was considerably shorter.

As mentioned earlier the acid production of Lb. casei is slower than that of the traditional lactobacilli in emmenthal cheese starters, and this means an accelerated ripening of the cheese as has been shown in the test cheeses. In this study we have succeeded in isolating Lb. casei-strains, combining new active starters, improving the aroma of emmenthal cheese and controlling its ripening. The results obtained can be regarded as notworthy in future developmental work.

REFERENCES:

Accolas, J.P. & Auclair, J. 1983 'Thermophilic lactic starters. Irish J. Food Sci and Technol. 7 (1) 27-38.

Langsrud, T. & Reinbold G.W. 1973 'Flavor development and microbiology of Swiss cheese - a review. II Starters, manufacturing process and procedures' J. Milk Food Technol., 36 (11) 531-535.

Steffen, C. 1982. 'Die Käseforschung im Dienste der Qualitätsförderung der Emmenthal-, Tilsiter- und Appenzellerkäserei' Schweizerische Milchzeitung 108 (94) 579-580

Thomas, T.D. 1985. 'Role of lactic acid bacteria and their improvement for production of better fermented animal products. A review' New Zealand J. Dairy Sci and Technol. 20 (1) 1-10

USE OF LYSOZYME FOR THE PREVENTION OF BUTYRIC ACID FERMENTATION IN GOUDA CHEESE. LIMITED EFFECT OF THE ENZYME*

J. Stadhouders & G. van den Berg
Netherlands Institute for Dairy Research (NIZO),
P.O. Box 20, 6710 BA EDE, The Netherlands

Lysozyme (500 U/ml cheese milk) can only prevent BAF if the number of spores of butyric acid bacteria (BAB) does not exceed 300 per litre (normal population of milk). In cheese from winter milk ($\geqslant$ 1000 spores/ l) lysozyme is therefore not effective. The phenomenon is explained by the presence of spores less sensitive to lysozyme. Cheese experiments with arbitrary pure cultures of BAB only, do not allow conclusions about the effectivity of lysozyme.

TUESDAY – POSTER 16

EFFECTS OF TEMPERATURE, LACTIC ACID AND SALT ON GROWTH OF AND ACID PRODUCTION BY THE MESOPHILIC MIXED-STRAIN STARTER BOS*

M. Northolt and G. Otten
Netherlands Institute for Dairy Research,
P.O. Box 20, 6710 BA EDE, The Netherlands

The influence of temperature, pH, lactate and NaCl on the growth of and acid production by a starter was studied. Conditions in Gouda cheese were simulated in growing and stationary cultures. The maximal growth rate was measured at 30°C, whereas the maximal acid production rate per bacterium (R_L) took place at 38-42°C. Growth was strongly inhibited at pH $<$5.7 and lactate concentration $>$1.1%, while also R_L decreased. R_L was strongly inhibited by NaCl.

*Complete poster available from authors.

QUANTIFICATION OF CHYMOSIN AND PEPSIN IN BOVINE AND CALF RENNETS

S. Visser, M.K. Friedenthal[1)] and H.S. Rollema
Netherlands Institute for Dairy Research, P.O. Box 20,
6710 BA EDE, The Netherlands
1) Tallinn Technical University, Tallinn, Estonia (USSR)

For the quantitative determination of chymosin and pepsin in calf rennet and adult bovine pepsin A the method of Garnot et al. (1) has been adopted now as a provisional IDF standard procedure (2). In the first step of this method the rennet sample is desalted after which both enzymes are separated by chromatography on a DEAE-column. In a second step the clotting activity of each of the separated enzymes is determined using a reconstituted milk substrate. Results are then expressed as clotting activity units. An extension of this method was obtained by the standardization of the milk substrate using reference enzymes which were themselves standardized on a well-defined hexapeptide substrate. With this extension clotting activities are converted into weight amounts of chymosin and bovine pepsin (3). Other quantitative determinations are based on immunochemical methods or on coagulation assays carried out at two different pH-values (4).

In this communication we present a relatively rapid determination procedure which does not require the use of specific antibodies nor necessitates desalting and/or chromatographic separation of the enzymes. Our method includes the specific inactivation of chymosin (15 min) in a 4.6 M urea-containing buffer at pH 2.0 and 30 °C (5). Before and after inactivation the proteolytic activity of the preparation towards the synthetic hexapeptide Leu-Ser-Phe-Nle-Ala-IleOMe is measured spectrophotometrically at 230 nm under specified conditions (6). The activities thus determined for the two enzymes are then converted to weight amounts by using specific proteolytic activities derived from experiments with highly purified chymosin and bovine pepsin A under the same conditions. Using specific clotting activities (7) the weight amounts of both enzymes could be converted further to milk-clotting activities.

For self-prepared mixtures of the purified enzymes the method gave excellent recoveries, whereas for commercial rennets reproducible results were obtained (coefficient of variation about 5 %).

References

1. P. Garnot, J.L. Thapon, C.M. Mathieu, J.L. Maubois and B. Ribadeau Dumas J. Dairy Sci. 55(1972) 1641-1650.
2. Provisional International IDF Standard 110: 1982.
3. P. Martin, J.C. Collin, P. Garnot, B. Ribadeau Dumas and G. Mocquot, J. Dairy Res. 48(1981) 447-456.
4. G.A.L. Rothe, N.H. Axelsen, P. Jøhnk and B. Foltmann, J. Dairy Res. 43(1976) 85-95.
5. D.M. Mulvihill and P.F. Fox, J. Dairy Res. 44(1977) 319-324.
6. S. Visser and H.S. Rollema, Anal. Biochem. 153(1986) 235-241.
7. G.A.L. Rothe, M.K. Harboe and S.C. Martiny, J. Dairy Res. 44(1977) 73-77

COMPLEXITY OF THE NATIVE CELL WALL PROTEINASE OF STREPTOCOCCUS CREMORIS HP AND PURIFICATION OF THE ENZYME

F.A. Exterkate and G.J.C.M. de Veer
Netherlands Institute for Dairy Research
P.O. Box 20, 6710 BA EDE, The Netherlands

The native cell wall-associated proteinase from Streptococcus cremoris HP is a high molecular weight complex structure. Two electrophoretically distinguishable proteolytically active components (1 and 2) were released from this organism when milk-grown cells were resuspended in a calcium-free buffer. Both components may disintegrate to give several catalytically active, partly less stable components of distinguishable, lower molecular weight. Component 2 (Mr 125 000) appeared to be a relatively more stable derivative of component 1 (Mr 135 000). The smallest proteolytically active derivative identified has a molecular weight of approx. 60 000. The proteinase has been purified with acceptable yields (total recovery of activity about 46 %), albeit mainly as component 2, using the affinity adsorbent carbobenzoxy-D-phenylalanine-triethylene-tetramine-sepharose. A similar result with respect to the derivative isolated and the efficiency of the procedure was obtained by means of co-precipitation of the enzyme during dialysis followed by sepharose CL-6B gelfiltration (total recovery of activity about 52 %). Gelfiltration of an ammoniumsulphate fraction appeared to be appropriate to isolate mainly component 1.

The enzyme is a serine proteinase with a pH optimum around 6.4 if determined at 30 ^{o}C and with [^{14}C CH_3] β-casein as the substrate. The optimum termperature was at 40 ^{o}C in the presence of Ca^{2+}-ions and at pH 6.2. The action of the enzyme has been established to be directed preferably to β-casein and is characterized by a relatively fast initial progress of the degradation. Ca^{2+}-(and also Mn^{2+}-) ions confer stability on the enzyme. Bivalent cations (possibly Ca^{2+}) appear to have an essential structural function so far that they maintain the catalytically (most efficient) conformation. The results suggest stabilization of the native proteinase in a rather complex structure of catalytically active units which is integrated within the cell wall polymer network.

pH-INDUCED PHYSICO-CHEMICAL CHANGES OF CASEIN MICELLES IN MILK AND THEIR EFFECT ON RENNETING

A.C.M. van Hooydonk, H.G. Hagedoorn and I.J. Boerrigter
Netherlands Institute for Dairy Research, P.O. Box 20,
6710 BA EDE, The Netherlands

Skim milk and rennet-treated skim milk were found to show a different pH-dependence of the dissociation of casein from the micel and the micel voluminocity. At pH 6.0 the rate of the enzymic reaction is at a maximum; the voluminocity then has a minimum value. The gelstrength of a rennet curd reaches a minimum at pH 5.3. This coincides with a peak value of the voluminocity.

PROPIONBAKTERIEN BEI DER HERSTELLUNG DER KÄSE VOM EMMENTALERTYP

J.Jelínek, P.Březina, L.Čurda
Chemisch-Technologische Hochschule, Prag, Tschechoslowakei

In der Arbeit wurde die biochemische Aktivität von vier gewählten Stämmen der Propionbakterien untersucht, die in der Milchtechnologie in der ČSSR verwendet werden: P.freudenreichii, Stamm Laktoflora 164, P.freudenreichii, subsp. shermanii, Stamm Laktoflora 160, Schweizer Propionkultur, Stamm Laktoflora 336 und P.freudenreichii, subsp. shermanii, Stamm Lyon 2. Die Aktivität wurde aus dem Gesichtspunkt ihrer Wachstums- und Gärungsaktivität, der CO_2--Bildung, der Bildung von flüchtigen Fettsäuren und der Produktion von Vitamin B_{12} verfolgt.

Die untersuchten Stämme werden bei der Versuchsproduktion der Emmentalerkäse in zwei Molkereien in der ČSSR eingesetzt. Die Herstellung und der Reifungsproze3 fanden in allen Fällen unter gleichen Bedingungen statt. Es wurde der Verlauf der Propiongärung verfolgt und gro3e Aufmerksamkeit wurde der sensorischen Beurteilung gewidmet.

Bei dem Reifen der Versuchskäse wurde die höchste biochemische Aktivität bei der Kultur P.freudenreichii, subsp. shermanii, Stamm Lyon 2 festgestellt, die eine grö3ere Menge von freien Aminosäuren vor allem Prolin bildete, im Vorlauf der Propiongärung mehr Propionsäure produzierte und an der Proteolyse der untersuchten Käse am stärksten von allen untersuchten Propionbakterien-Stämmen beteiligt war. Dieses Ergebnis wurde ebenfalls durch sensorische Analyse nachgewiesen.

Die Arbeit hat nachgewiesen, da3 die Käse Qualität der Emmentalerkäse durch Verwendung von verschiedenen Propionbakterien bedeutend beeinflu3t werden kann. Ebenfalls durch die Auswahl von geeigneten Stämmen kann die Reifungsdauer reguliert werden, sie kann verkürzt oder verlängert werden, wobei die sensorischen Eigenschaften erhalten werden. Aus diesen Gründen ist es unbedingt nötig, die biochemische Aktivität der Propionbakterien-Stämme in der Technologie der Herstellung der Emmentalerkäse gründlich zu kennen.

PROTEINASE UND PEPTIDASE DER BREVIBACTERIUM LINENS: EIGENSCHAFTEN UND DEGRADATION DER BITTEREN PEPTIDE IN DER KÄSEN

P.Březina, J.Kopečný[x], M.Plocková, I.Schovánková
Chemisch-Technologische Hochschule, Prag,Tschechoslowakei
x Institut für Physiologie und Genetik der Wirtschaftstiere ČSAV, Prag, Tschechoslowakei

Der bittere Geschmack der Käse stellt zur Zeit ein ernstes Problem der Technologie der Käseerzeugung vor. Zur Behebung dieses Geschmacks können Enzympräparate benutzt werden. Einer der geeigneten Produzenten solcher Enzymen ist Brevibacterium linens.

In unseren Versuchen wurden im Ganzen 12 Stämme B.linens testiert /unsere Isolaten und Isolaten aus der Europ.sammlungen der Mikroorganismen/. Bei diesen Stämmen wurden folgende Wachstumsgeschwindigkeiten, der Proteinase- und Aminopeptidaseproduktion. Für weitere Arbeit wurden vier folgende Isolaten ausgewählt: JH, a, L01, 200, die die höchste Menge der Extrazelularproteinase und Aminopeptidase produzierten. Beim Stamm 200 wurden die Eigenschaften der erwählten Enzymen enger beschrieben. Diese Enzymen wurden mit der Hilfe Gellfiltration und Affinitätschromatographie auf HMDA-Zellulose isoliert. Das gewonnene Präparat wurde mit der Hilfe der Elektroforese auf SDS-PAG charakterisiert. Optimale pH Proteinaseund Aminopeptidase war 8,0 bzw. 7,8, die Optimaltemperatur beträgt 52 °C, bzw. 32 °C.

Die Proteinase sind überwiegend des Serintyps und Aminopeptidase sind typische Metalopeptidase durch Kobalt aktiviert.

Extrazelularenzymen, gewonnene aus den Stämmen JH, a, L01, 200, wurden zur Verfolgung der Möglichkeit der Degradation der bitteren Peptide benutzt. Diese wurde aus den Käsen des Emmentalertyps mit erhöhten Auftreten des bitteren Geschmacks und zwar durch die Extraktion und Verteilung durch Gellfiltration gewonnen. Die gewonnenen Aufschlußgeschwindigkeiten der bitteren Peptide bringen die Voraufsetsungen für ihre Testierung direkt bei der Käseproduktion.

New applications of rennet mixtures in cheesemaking

Hušek,V.,Minařík,R.,Součková,M.
Dairy Research Institute, Českobrodská 1174, 198 00 Praha 9 - Kyje
CSSR

As the result of laboratory tests we suggested rennet mixtures containing chicken pepsin (CP),Mucor miehei rennet (MM) in ratio of milk clotting activity 50/50 or 60/40,chicken,bovine (BP) and swine (SP) pepsin in ratio 50/30/20. These mixtures were assessed for Edam type cheese production on the plant scale. The result obtained were compared with cheesemaking experiments with other mixtures of rennet without chicken pepsin as well as with sole coagulant. During these comparative experiments were used mixtures of MM and SP 50/50,MM,BP and SP in ratio 40/40/20 or sole coagulants i.e., commercial chymosine rennet (CH), chymosine rennet obtained by fistulating suckling calves and MM rennet. Some differences could be detected in consistency of the experimental cheese made with CP in rennet mixtures. The softer body of these cheese and changes in rigidity, elasticity and viscosity rather diminished with time of cheese maturation. The ratio of soluole and aminoacid nitrogen to the total nitrogen in experimental cheese is related to the type of rennets used. Especialy those cheese produced with rennet mixtures containing CP showed higher ratios both of soluble and aminoacid nitrogen. Cheese made with rennet mixtures composed by CP showěd also slightly more pronounced flavour when compared with control cheese and resembled more matured cheese. The commercial production of Edam type cheese and Moravian loaf and block cheese with the mixtures of rennet containing 30-50% of chicken pepsin from the whole milk clotting activity is now well adopted by the cheese industry of CSR.

Table 1 The ripening of Edam cheese after 2 months

Exp.No	Rennet type	MCA (%)	DM (%)	pH	SH	TN (%)	SN (%)	AN (%)
1	CH	100	57,14	5,62	83	3,91	18,13	9,33
2	"in vivo"	100	58,36	5,40	87	3,87	16,47	8,88
3	MM	100	58,60	5,58	83	3,92	17,70	11,34
4	CP,MM	50/50	57,55	5,50	84	4,00	22,66	12,73
5	SP,MM	50/50	57,34	5,52	85	3,35	18,14	10,74
6	CP,BP,SP	50/30/20	57,43	5,65	88	3,92	16,24	13,52
7	MM,BP,SP	40/40/20	57,77	5,62	87	3,82	15,64	12,29

MCA-milk clotting activity,DM-dry matter,pH of 10g cheese slurry+30 ml water,SH -acidity of 10g cheese+20 ml water in ml 0,25mol NaOH per 100g cheese,TN-total nitrogen,SN-soluble,AN-aminoacid nitrogen as % of TN,in vivo-chymosin from fistulated calves

The research of the new sorts of starter cultures for cheese manufacture

Minařík R.
Dairy Research Institut Tábor, Czechoslovakia

The starter cultures having the antagonistic effect against the undesirable bacteria are applied in cheese making sincle lately thus improving the quality of cheese. We have studied the application of cultures L. Plantarum as well as L. Lactis, the latter of which has the higher producion of lactic acid as well as the better premises of its industrial application in comparison with L. Plantarum.
The six strains of L. Lactis from the Collection of Dairy Industri in Prague (Czechoslovakia) have been tested. All tested strains inhibited the coliformes as well as the spore-forming bacteria. The trials have been carried out on a comprehensive large scale, this communication however is limited only to the obtained practical results.
The cultures L. Lactis have been applied in a volume of 0.02 % besides of the current mesophilic bacteria in the manufacture of various Dutch-type cheese. The way of cheese processing was not aletered. Cheeses ware pressed one hour as a rule and they were salted immediately.
The application of L. Lactis led to a somewhat more rapid lactic acid production in the first stage of the process, to a bit higher solids content and to a higher acidity before salting respectively.
During the course of the ripening a somewhat higher but regular breakdown of protein occured. The limitation of bacteria undesirable from the stand-point of cheese processing has led to the better structure, eyesfromation, cleaner and more pronounced flavour and to the prolonged keeping quality of produced cheese respectively.
The application of L. Lactis in Dutch-type cheese manufacture makes it possible also to eliminate the use of KNO_3.
The culture L. Lactis has been applied successfully as a component of a special mixed bacterial starter in the emmentaler type cheese manufacture. In the cheese process applying the mentioned culture, there is possible to salt the cheese immediately after six hour's pressing.

A Diffusion Model for Describing Sodium Chloride Movement During Salting of Cheese in Brine

Josef Horký, Prokop Nekovář[+], Pavel Březina[+]
General Management, Dairy Industry, Prague, Czechoslovakia
[+]Institute of Chemical Technology, Prague, Czechoslovakia

The diffusion of NaCl into the surface zone of cheese was studied during brine salting. The content of NaCl in slices of about 1-3 mm thickness scraped off was determined by the ion chloride selective electrode method. Measurments were performed with model cheeses and brines, the composition of which was exactly defined. The results of these experiments were proved by measuring in cheese factories.

The determination of an acceptable methematical model of NaCl diffusion was conditioned by presumptions such as: the constant value of the diffusion coefficient of NaCl during brine salting, the constant concentration of NaCl in brine (x_L), the perfect stirring of brine, the symetric one-dimensional mass transport, the homogenic structure of cheese curd, the well-known equilibrium concentration values of NaCl on the boundary line cheese-brine.

The diffusion coefficient values of NaCl in cheese were determined by two methods based on the Fick's second law:

1. the method using penetration of NaCl through a cheese slice fixed into the diffusion cell apparatus in which the space with brine is separated from the distilled water by the cheese slice;
2. measurments of NaCl diffusion between two compact blocks of cheeses, the former was salted, the latter was not, both were lying contacting each other. The equilibrium values of concentration of NaCl (x_R) on the boundary line cheese-brine were obtained by statistic treatment of dates after a long time of salting. The relationship between x_R and x_L was found: $x_R = q.x_L + k.x_L^2$, k and q are characteristic constants for each type of cheese (dry matter) and the temperature of brine. Both different methods provided practically coincident results for several types of cheeses.

A mathematical model was proposed to describe the intake of NaCl into the cheese surface during brine salting according to the above-mentioned conditions. It could be seen that theoretical diffusion curves simulated by computers fitted the experimental values very well indicating the accuracy by wicht the proposed diffusion model could predict the NaCl intake into the cheese surface during brine salting.

THE EFFECT OF MILK PROTEIN GENETIC VARIANTS ON CHEESEMAKING

E.R.B.GRAHAM, D.M.MCLEAN, P.ZVIEDRANS

INTRODUCTION

The composition and cheesemaking properties of milk are affected by the milk protein genetic variants of the cow. Milk containing the type B genetic variants of β-casein and κ-casein have a shorter rennet coagulation time and give a firmer curd than the A variants. The type B variant of β-lactoglobulin is associated with a higher concentration of casein in the milk and a higher proportion of casein in the milk protein.

In this experiment cheddar cheese was made from milk of individual friesian cows having all type A genetic variants of β-casein, κ-casein and β-lactoglobulin or predominantly type B genetic variants of the same proteins.

METHOD

Type B cows were paired within friesian herds with A cows having similar stage of lactation, age and milk production. Only milk with a somatic cellcount of less than 750,000 cells per ml was taken for cheesemaking. The milk was analysed for fat, protein and casein and standardised to a casein to fat ratio of 0.68. Each milk was tested for its rate of development of (i) curd firmness on addition of rennet and (ii) acid production on addition of different amounts of frozen direct set starter. The time of rennet addition and the amount of starter added were adjusted for each pair of milks so that the rennet curds were cut at the same time and the rates of acid production were similar.

Cheddar cheese was made using traditional methods from 12 kg amounts of milk from individual cows and from 200 kg amounts of milk which had been pooled according to genotype. The cheese was analysed ex-press.

CHEESEMAKING VATS

RESULTS

The time taken for the A milk to develop the curd firmness required for cutting was at least 50% (15 minutes) longer than that for the B milk (Table 1). The yield of cheese solids per unit of protein in the milk was more than 5% greater for the type B milk compared to that of the type A milk both for individual cow milks and for pooled cow milk. Cheese solids was significantly related to the casein concentration in the milk. The type B milk retained more fat and protein in the cheese than the type A milk (Tables 1 and 2).

TABLE 1

MEAN YIELD OF CHEDDAR CHEESE MADE FROM 12 KG MILK FROM INDIVIDUAL COWS HAVE TYPE A AND TYPE B GENETIC VARIANTS OF THREE MILK PROTEINS

	A	B	100B/A
NO OF COWS	14	14	
MILK PROTEIN, %	3.01	2.97	98.7 (n.s.)
TIME TO CUT, MIN	47	30	157 (P<.001)
CHEESE SOLIDS, g	714 ± 21*	741 ± 20*	103.7 (n.s.)
CHEESE SOLIDS/PROTEIN	2.38	2.51	105.1 (P<.001)
FAT RECOVERY, %	88.9	91.5	103.0 (P<.001)
PROTEIN RECOVERY, %	76 8	79.2	103.7 (P<.001)

TABLE 2

COMPOSITION OF CHEDDAR CHEESE MADE FROM 12 KG MILK OF TYPE A AND TYPE B GENETIC VARIANTS OF THREE MILK PROTEINS

	A	B
MOISTURE, %	35 8	35.6
PROTEIN, %	24.7	24.5
FAT, %	33.0	33.9 (P<0.01)
FAT IN MOISTURE - FREE-SOLIDS, %	51.4	52 5 (P< 01)

CONCLUSION

Friesian cows having type B genetic variants of β-casein, κ-casein and β-lactoglobulin produce superior milk for cheesemaking due to a shorter manufacture time and a higher yield of cheese. Therefore increasing the proportion of type B genetic variants in the dairy cow population would improve the profitability of cheese manufacture. It would not be difficult to include milk protein genetic variants as an additional criterion in the breeding of dairy cattle by selecting type B bulls for artificial breeding.

NORTHFIELD RESEARCH CENTRE, DEPARTMENT OF AGRICULTURE, ADELAIDE, SOUTH AUSTRALIA

INFLUENCE OF A PRE-ACIDIFICATION IN MAKING SAINT-PAULIN BY FULL-CONCENTRATION WITH ULTRAFILTRATION

R. DELBEKE, G. WAES, R. VAN RENTERGHEM, M. NAUDTS

Ultrafiltrationretentate		Cheese
46 % total solids (TS)	additives, treatment no whey expulsion	50 % TS
21 % proteins (P)	→	23 % P
21 % fat (F)	4 weeks ripening	23 % F

To avoid

- strong acidity, unpleasant deteriorated taste, lack of taste
- mealy texture, lack of elasticity

A combination is needed of

- retentate with less than 1.9 % lactose and 1.9 % ash
- pre-acidification for ca. 16 h at 20°C
- renneting at pH 5.2 - 5.0

Government Dairy Research Station, 9230 Melle, Belgium

FRACTURE PROPERTIES OF GOUDA CHEESE

General

Fracture is initiated if.

Stress is higher than the adhesion-cohesion forces in the material. Inhomogeneities in the material cause stress concentration, which causes fracture to occur at a lower overall stress than would otherwise be the case.

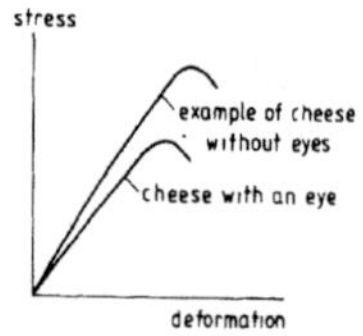

An initially formed crack then spontaneously propagates as long as the strain energy released by progress of a crack suffices to provide the energy needed to create new crack surfaces.

Another important parameter is the deformation at which fracture occurs (= εf).

The fracture stress (= σf) proved to be independent of the type of deformation, the overall deformation at fracture was not.

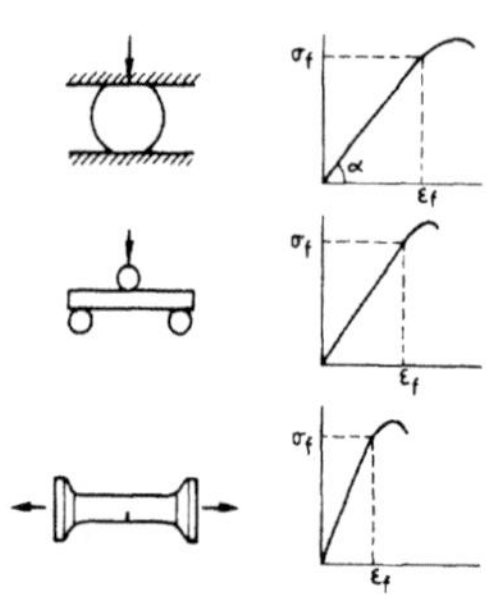

By applying different size notches in a sample tested in tension, additional information about the fracture behaviour can be obtained.

Some results

The magnitude of the fracture parameters depended on:

- The rate of deformation $\dot{\varepsilon}$.

Cheese is a visco-elastic material. This means that part of the strain energy is dissipated during the deformation, which part depends among others on the deformation rate. The results were indeed found to be depended on the deformation rate.

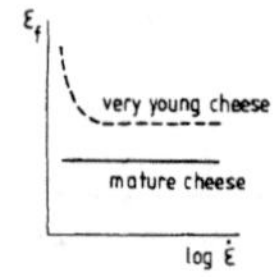

- Moisture and fat of cheese.

A cheese with a lower moisture content generally is firmer. This can be measured by means of the initial slope (tgα) of the deformation curve, which is a kind of complex modulus. This modulus however, may or may not be correlated with the fracture parameters. Fracture parameters markedly depend on fat content and on maturation.

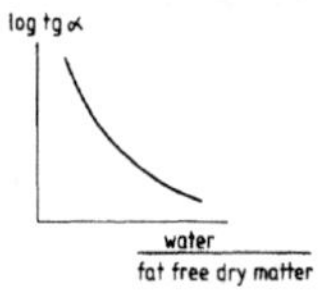

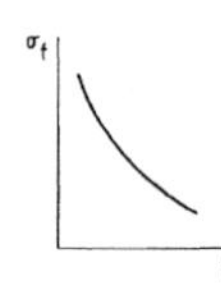

- Maturation and proteolysis of cheese

Maturation primary implies proteolysis, but it commonly goes along with a decrease in moisture content. Consistently, the deformation needed for fracture decreased with advancing maturation, although during the first month the decrease was marginal.

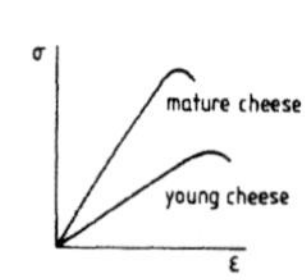

lh
Agricultural University
Dept. of Food Science
Biotechnion, De Dreijen 12
6703 BC Wageningen, The Netherlands

H. Luyten
T van Vliet
P. Walstra

EFFECT OF MECHANICAL PRESSURE ON THE SYNERESIS OF RENNET CURD

INTRODUCTION

Syneresis rate is greatly influenced by process conditions, such as pH, temperature and mechanical treatment. To enhance our understanding of syneresis of individual curd particles, a one-dimensional approach was used (1). In this way experimental results (syneresis of thin slabs of curd) could be compared with model calculations. Syneresis can be described by the equation of Darcy, which states that the linear flow rate v is given by

$$v = B \Delta P / \eta \, l \qquad [1]$$

where B is the permeability coefficient, η is the viscosity of the whey and $\Delta P / l$ is the pressure gradient. The pressure exerted on the whey ΔP may result from:

1. Endogenous syneresis pressure due to the rearrangement of the casein network (0 - 3 Pa).
2. Mechanical pressure due to application of a load on the slab (0 - 60 Pa).
3. Pressure due to gravitational force (0 - 1 Pa).

During actual cheesemaking the endogenous pressure and the variable mechanical pressure, caused by cutting, stirring and pressing, seem to be important. B, ΔP and l vary with the state of shrinkage, location and time. Applying equation [1] to small volume elements in combination with the equation of continuity gives an expression for syneresis rate, which can be numerically solved (2).

METHOD

The microscope method used by Van Dijk (1) for measuring the shrinkage of thin slabs of curd was extended, as outlined in Figure 1. After starting syneresis by

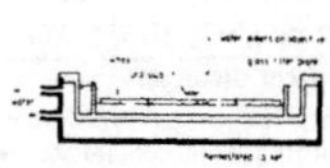

moistening the surface of the slab, mechanical pressure was applied, using highly permeable glass filter plates. The shrinkage was determined by focusing on the surface of the glass filter plate and reading off the change in height on the micrometer. Reconstituted skimmilk was used, pH adjusted by HCl, with 500 ppm rennet at 30 °C. Syneresis was started 30 minutes after rennet addition, while mechanical pressure was applied 60 seconds thereafter.

RESULTS

Figure 2 shows the initial shrinkage of the slab to be proportional to $\sqrt{\text{time}}$ after application of the extra load, in accordance with theories for consolidation processes. Results of model calculations of the shrinkage 5 minutes after starting syneresis were in reasonable agreement

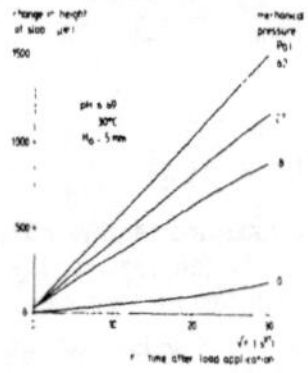

with the experimental results: Figure 3. Deviations may be due to the visco-elastic character of milk gels (3). Further studies concerning the influence of rheological properties on the effect of mechanical pressure are in progress.

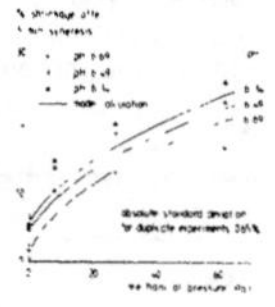

CONCLUSIONS

1. The influence of mechanical pressure on syneresis rate can be accurately determined.
2. A small mechanical pressure already leads to an overwhelming increase in the syneresis rate of rennet-induced milk gels in the initial phase of the syneresis process.
3. Application of consolidation theories may be useful in describing syneresis of milk gels.

Agricultural University
Department of Food Science
Laboratory of Dairying and Food Physics
Biotechnion, De Dreijen 12
6703 BC Wageningen, The Netherlands

REFERENCES

1. H.J.M. van Dijk, Syneresis of curd. Doctoral Thesis, Agricultural University, Wageningen, 1982.
2. H.J.M. van Dijk, P. Walstra & J. Schenk, Chem Eng. J. 28(1984)B43.
3. H. Kamphuis, R.J.J. Jongschaap & P. Bouter, J. Colloid Interf. Sci. 98(1984)459.

We are indepted to the Stichting J. Mesdagfonds for financial support.

H.J.C.M. van den Bijgaart
P. Walstra

PHYSICAL AND CHEMICAL CHARACTERISTICS OF GOAT MILK FROM YUGOSLAVIA AND ITS CAPABILITY OF COAGULATION

LJ. KRŠEV, Z. RAC*, LJ. TRATNIK
Faculty of Food Science and Biotechnology, Zagreb
*Dairy factory, Zagreb, "Dukat"

Introduction and Purpose of Research

To produce the goat cheese of good and unified quality, one should know physical, chemical and technical characteristics of milk, its capability of coagulation and characteristics of the coagulum.

The purpose of this work was to research the characteristics of goat milk from Yugoslavia, its capability of coagulation by microbial protease and then characteristics of gel.

Material and Method of Work

During the lactation in the year 1985., every ten days have been taken samples of goat milk (Alps and Sana) from herds which are raised in the north-west part of Yugoslavia (A) and from herds in the central part of Yugoslavia (B).

For coagulation have been used microbial protease : "Renilase","Fromase" and "Mikrozyme"

Analytic Methods

Analysis of total proteins, caseins, lactose and dry matters has been fixed by IDF regulations. Quantity of ach and Ca has been fixed by "British Standard" 1741/1963, milk fat by Gerber, acidity by "Soxsket-Henkel method, α_{s1}, α_{s2}, ß, K have been determined by electrophoresis (3-8,6 pH) toward Boulanger, α-lactalbumin and ß-lactoglobulin by electrophoresis of curdle (pH 8,6), hydratisation of micelles have been determined by measuring the humidity of sediment after centrifuge (65000 g), colloid Ca has been determined fluorometrically and total colloid inorganic P by colorimeter. Coagulation and gel characteristics have been researched by automatic torsional viscosimeter Plint.

Results of Work

Characteristics of goat milk are presented in tables 1 and 2.

Table 1.

		Middle value		Standard deviation		t-Test
		A	B	A	B	
Milk fat	(%)	4,40	4,8	0,200	0,310	5,891
Total proteins	(%)	3,3	3,45	0,130	0,223	0,582
Casein	(%)	2,53	2,61	0,094	0,235	0,677
Whey protein	(%)	0,83	0,91	0,044	0,09	1,811
Lactose	(%)	4,12	4,35	0,089	0,175	1,391
Ash	(%)	0,75	0,85	0,272	0,317	3,299
Ca	(%)	0,13	0,12	0,031	0,0	–
Acidity		0,15	0,16	0,0	0,0	–
pH		6,4	6,65	0,0	0,071	1,373
Density		1,0306	1,0308	0,0	0,0	–

$n_A = 35$ $n_B = 33$

Table 2.

	Middle value		
	Goat milk		Cow milk*
	A	B	
Casein α_{s1}	4,9	5,1	38
Casein α_{s2}	15,3	11,4	12
Casein ß	47,8	44,2	36
Casein K	18,5	16,8	14
α-lactalbumin/ß-lactoglobulin	0,58	0,62	0,4
Colloid Ca (% of total Ca)	61,3	62,8	65
Inorganic P (g/l)	0,66	-	-
Colloid P (% of total P)	50	-	-
Hydratisation of micelles (g/gDM)	1,62	1,7	1,9

* Informations from literature

Table 3 presents the coagulation capability of goat milk by microbial protease in comparison with coagulation by animal protease.

Table 3.

	Middle value for goat milk (A+B)				Middle value for cow milk
	Renilase	Fromase	Mikrozyme	Animal protease	Animal protease
Coagulation time (min)	18,7	19,5	19,0	19,6	35
Coagulum firmness (mV)	44,5	42,2	41,8	43,9	75

CONCLUSION

- It has been established relative coincidence of goat milk structure in two hested regions. Significant difference has been established for fat and ash
- Values for Ca, P and value for hydratisation of micelle are identical for both regions.
- In comparison with cow milk, in goat milk have been established smaller quantities of α_{s1} casein and larger quantities of α_{s2}, ß and K casein
- Hydratisation of micelle is larger in goat milk
- Floculation time and curdle strenght time are very different with microbial / animal.

ULTRAFILTRATION OF CHEESE CAMEMBERT BRINE

LJ. KRŠEV, Z. KORITNIK,* LJ. TRATNIK
Faculty of Food Science and Biotechnology, Zagreb
*Dairy factory, Zagreb, "Dukat"

Introduction and Purpose of Work

During its usage, the brine becomes "enriched" with mineral and organic matters and with microorganisms, which influences on cheese quality.

The purpose of work was to compare the application of ultrafiltration with the application of classical method of brine cleansing.

Material of Work

The brine samples are taken from industrial production of Camembert cheese. Every two days NaCl is added in brine because of density; brine used up to 14th day.

Methods of Work

Classical method of brine cleansing includes neutralisation and thermal treatment (95°C/3 min).

Ultrafiltration of brine is carried out throught DDS membrane-generation II. Total N is determined by micro-Kjeldalh method, mineral structure and lactose are determined by spectrophotometrical method

IDF regulations are used for microbiological analysis of samples.

Results of Work

Table 1 presents results of following the changes in chemical structure of brine.

Table 1
Results of Following the Change in Chemical Structure of Brine

Time days	1	3	5	7	9	11	13
Dry matter (g/l)	270,95	249,81	270,98	273,5	271,85	270,95	273,15
Density -	1,185	1,175	1,173	1,172	1,175	1,172	1,171
pH	7,1	5,5	5,1	4,82	4,68	4,6	4,55
NaCl (g/l)	270,77	246,6	264,95	260,11	258,13	257,15	258,25
Lactose (g/l)	0	0,65	1,31	1,42	1,95	2,31	2,41
Total N (mg/l)	0	58,6	92,8	136,6	161,38	171,85	183,1
Aprotein nitrogen (mg/l)	0	45,1	77,1	94,3	119,4	129,9	139,2
Ca (mg/l)	55	221	398	462	601	618	631
Mg (mg/l)	4	11	31	29	29	28	27

n = 5

Table 2 presents average microbiological structure of brine (first day and after 13th day)

Table 2. Average Microbiological Structure of Brine (in 1ml/Brine)

	First day	13th day
Mesophil Aerobic Microorganisms	1200	230000
Coli Bacteria	+	+
Micrococcus and Staphylococcus	50	12000
Yeast	3	51000
Molds	8	18000

Table 3 presents changes in chemical structure of brine after using the classical method and ultrafiltration

Table 3. Change in Chemical Structure of Brine After Using the Classical Method and Ultrafiltration

	Before treatment	After treatment	
		Classical (thermal treatment)	UF method
pH	4,56	6,7	4,4
Total N (mg/l)	198,41	147,20	130,12
NPN (mg/l)	140,9	123,5	118,91
Ca (mg/l)	64,1	85,0	270,0
Mg (mg/l)	26	10	17
Lactose (g/l)	2,38	1,69	1,38
Sediment (g/l)		10	

Table 4 contains the informations about microbiological structure or brine after thermal treatment and ultrafiltration

Table 4. Microbiological Structure of Brine After Classical Method and Ultrafiltration (in 1 ml/Brine)

	After thermal treatment	After UF method
Mesophil aerobic microorganisms	0	80000
Coli bacteria	-	-
Micrococcus and Staphylococcus	1	-
Yeast	0	2
Molds	0	0

Conclusion

- During usage of Camembert cheese brine, the quantity of organic and mineral matters and microorganisms is increasing in it, while pH is falling.
- Application of thermal treatment in cleansing the brine significantly reduces the quantity of organic and mineral matters and helps to attain a microbiological cleanlinees of brine
- Good microbiological cleanliness and noticeable demineralisation of brine are attained by ultrafiltration
- UF method is simple and economical and and doesn't cause corrosion, so that it is better treatment for cleansing the brine

Autolysis of lactic streptococci in milk

Gerd Vegarud and Thor Langsrud
Department of Dairy and Food Industries,
Agricultural University of Norway,
N-1432 Aas-NLH, Norway

INTRODUCTION

Autolysis of lactic acid bacteria are important for cheese ripening, with release of peptide hydrolases.

Conditions favouring autolysis of the cells may accelerate the cheese ripening prosess.

The purpose of the present work was to study:

1. The rate and extend of autolysis and release of intracellular peptidases from some strains of group N streptococci in milk.

2. The influence of cheese rennet on the autolytic activity of the cells.

METHODS

- The bacterial strains used:
  - Streptococcus lactis L2 and L4
  - Streptococcus lactis subsp. diacetylactis ATCC 15346
  - Streptococcus cremoris C12
- All strains were grown in skim milk at constant pH for 20 days at 20°C or 30°C.
- Autolysis of the cells were measured by:
  - The change in absorbance at 480 nm (A_{480}) after solubilization of the casein micelles. □
  - The concentration of cellbound and released DNA. ● ○
  - The release of Lactate dehydrogenase (LDH) ■
  - The release of prolineimino peptidase (PIP) ▲

RESULTS

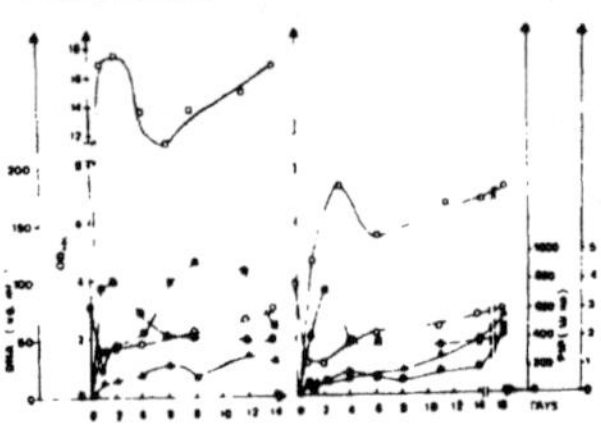

Growth and autolysis of S. lactis L2 in milk at 30°C without cheese rennet — with cheese rennet

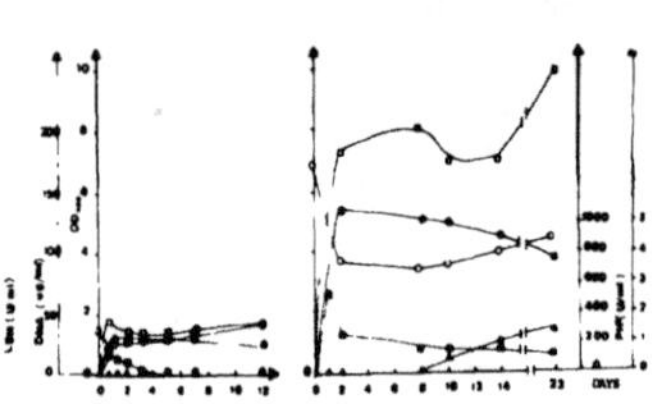

Growth and autolysis of S. lactis subsp. diacethylactis ATCC 15346 in milk at 30°C without cheese rennet — with cheese rennet

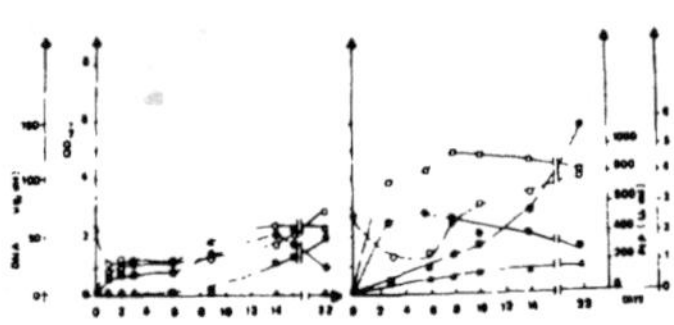

Growth and autolysis of S. cremoris C12 in milk at 20°C without cheese rennet — with cheese rennet

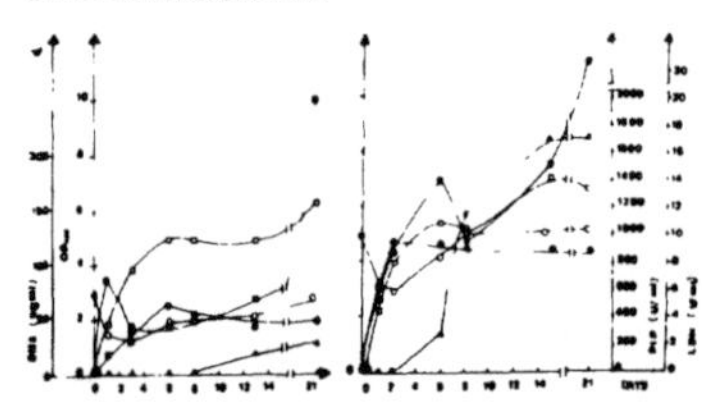

Growth and autolysis of S. lactis L4 in milk at 30°C without cheese rennet — with cheese rennet

CONCLUSION

- S. lactis L2 autolysed rapidly in skim milk at constant pH and 30°C.
- S. lactis subsp. diacetylactis ATCC 15346
  S. lactis L4
  S. cremoris C12 autolysed slowly
  (30°, 30°, 20°C)
- Cheese rennet stimulated the autolysis of
  S. lactis subsp. diacetylactis ATCC 15346
  S. lactis L4
  S. cremoris C12

ACCELERATED CHEESE RIPENING WITH DEBARYOMYCES HANSENII AS AN ADDITIONAL STARTER

P. Deiana, F. Fatichenti
Istituto di Microbiologia Agraria e Tecnica Università di Sassari-Italy.

In Pecorino Romano cheese-making trials on an industrial scale, a selected culture of Deb. hansenii was added to the lactic acid bacteria starter. The yeast developed well towards both the centre and crust of the cheese, and proteolysis was quicker. The production of volatile and non-volatile acids not only took place more quickly, but also proved to be more abundant in the interior of the wheels than towards the exterior. NCN and NPN were produced in greater quantities. As well as overall ripening being accelerated, superiority over the controls resulted in colour, texture and organoleptic properties.

TUESDAY - POSTER 33

A KINETIC STUDY OF OVOMUCIN INHIBITION ON CHYMOSIN

I. Söderberg
Department of Food Technology, University of Lund, P.O. Box 124, S-221 00 Lund, Sweden

Ovomucin, a glucoprotein, has earlier been described as an inhibitor of milk coagulation. As such it could be used to recover chymosin from whey. Therefore the inhibition of chymosin with ovomucin was studied with the commercial hexapeptide HLeu-Ser-Phe-(NO_2)-Nle-Ala-Leu-OMe as substrate.

MINERAL CONTENT OF FRESH AND SEMI-HARD GOAT'S CHEESE

M.C. Martín-Hernández and M. Juárez
Instituto del Frío (CSIC). Ciudad Universitaria.
28040 Madrid, Spain.

The main and trace mineral element contents of fresh and semi-hard goat's cheeses during manufacturing were studied.
The samples were taken from four batches of each type of goat's cheese prepared semi-industrially. Fresh cheeses were made without the addition of starter. The process for preparing the semi-hard cheeses, which made with starter, included separating a part of the whey and replacing it with the same quantity of a 5% NaCl solution.

Table 1. Distribution of mineral elements during the manufacture of fresh and semi-hard goat's cheese and retention in brined cheeses

	Fraction	Total solids (%)	Na	Ca	Mg	P	Fe	Cu	Zn	Mn
			(mg/100 g)				(mg/kg)			
Fresh	Milk	12.97	52.8	129	11.1	80.9	0.37	0.25	2.77	0.033
	Whey	-	62.7	47.3	8.3	34.7	0.13	0.07	0.27	0.007
	Curd	30.28	37.0	401	18.7	245	1.24	0.68	11.19	0.107
	Brined cheese	42.58	200	600	23.1	345	1.78	0.78	14.81	0.169
	Retention (%)	-	16.8*	79.1	35.4	72.5	81.8	52.9	90.9	87.1
Semi-hard	Milk	13.44	44.2	135	10.2	83.0	0.36	0.21	3.28	0.032
	Whey	-	45.0	50.0	7.0	32.0	0.11	0.06	0.12	0.008
	Curd	34.40	47.5	390	15.5	223	1.62	0.76	12.53	0.135
	Brined cheese	45.60	79.0	573	26.6	343	2.35	0.75	18.51	0.203
	Retention (%)	-	18.3*	55.2	33.9	53.7	84.9	46.4	73.4	82.5

* Calculated on a curd yield of 24% (fresh) and 17% (semi-hard).

The results indicate that, with the exception of Na and Mg, the mineral elements present in the milk are retained in the brined cheese in a proportion higher than 46%. Calcium was retained to a greater extent than the other major elements, followed by phosphorus. A substantial increase was observed in the trace mineral element content as compared to the trace element content of the milk.
The differences found between the two types of goat's cheese studied may be explained by the lower pH of the semi-hard cheese resulting from lactic fermentation.

CHANGES IN THE NITROGEN AND LIPID FRACTIONS OF CABRALES CHEESE

L. Alonso, M. Juárez and M. Ramos*
Instituto del Frío (CSIC). Ciudad Universitaria. 28040 Madrid, Spain.
*Instituto de Fermentaciones Industriales (CSIC)
Juan de la Cierva, 3, 28006 Madrid, Spain.

Cabrales cheese is artisanal blue cheese made from cow's milk with 20-30% goat's and ewe's milk. This study deals with the changes taking place in the proteins and lipids in Cabrales cheese during ripening.
Five batches of cheese were prepared. Analysis of the gross composition, nitrogen fractions, fat indices, free fatty acid and glyceridic fatty acid composition were carried out. Electrophoretic analysis of the caseins was also performed.

Mean values of compositional characteristics, soluble nitrogen (SN), non protein nitrogen (NPN), free amino acids (NH_2-N) and Reichert-Meissl (RM), Polenske (P), Kirschner (K), refraction (I_R^C), free fatty acid (FFA) for Cabrales cheese during ripening.

Aspect	Ripening time				
	2 days	1 month	2 months	3 months	4 months
pH	5.31	6.04	6.26	6.59	6.77
Fat (%)	29.25	35.31	33.75	32.88	30.63
TS (%)	59.27	69.23	66.07	63.69	61.25
NaCl (%)	2.08	2.87	3.81	3.69	3.73
SN (% TN)	19.10	35.70	58.06	81.74	88.09
NPN (% TN)	9.15	22.54	39.94	70.03	69.47
NH_2-N (% TN)	0.7	5.9	10.8	19.4	22.0
RM	26.2	24.2	20.9	19.3	18.2
P	5.3	4.3	4.0	3.4	3.2
K	23.2	20.7	19.6	17.5	14.9
I_R (40°C)	1.4532	1.4534	1.4511	1.4493	1.4478
FFA*	1.69	6.93	11.29	19.38	28.05

TS = Total solids; TN = Total nitrogen; *mg KOH/g fat

Extensive proteolysis and lipolysis occurred in this type of cheese. Approximately 88% of the total nitrogen in the mature cheese was water soluble, and almost 79% of the water-soluble nitrogen was non-protein nitrogen. Degradation of α_s - and β- casein was nearly complete. The free fatty acid content was high, and there was a significant decrease in the short chain acids of the glyceridic fraction.

CONTRIBUTIN OF S. LACTIS STRAINS AND STRAIN MICROCOCCUS M–104 ON CASEIN DEGRADATION

D. Obradović
University of Belgrade, Faculty of Agriculture
Departmant of Microbiology
Beograd, YUGOSLAVIA

It has been found that in conjoint cultures S. lactis AK–60 and Micrococcus M–104 grow normaly with mutualistic stimulative action. As a starter cultures in mixed populations, these strains contribute towards normal curdling process, and formation of the curd. This starter could significantly enhance the taste and aroma of the cheese (trapist, white cheese) as well as reduce the ripening time.

In this work the cells of these two strains and other two S. lactis strains have been ultrasonicaly disrupted, and cell fractions were tested for proteolytic activity against native casein. Polyacrylamide gel electrophoresis was used to study the breakdown of major casein fractions. Intracellular extracts and cell wall fractions of streptococci caused a very similar alteration in the electrophoretic behavior of the casein. Both γ *– and* β *– casein were found to be much more hydrolysed than* α_{S_1}*, specialy when the cell wall fractions have been tested. The most pronounced proteolytic activity is shown by Micrococcus M–104 cell wall which showed a great similarity to that of renin. That was documented with characteristic electrophoretic pattern. It could be that this cell wall associated proteinase activity serves a function in nitrogen metabolism, of the cells of conjoint cultures, hydrolysing large casein aggregates to smaller, which in turn could be further hydrolysed by streptococcal enzymes.*

The study on accelerated ripening period of sombor cheese in industrial conditions

D. Petrović, D. Mišić – University of Belgrade, Faculty of Agriculture, 11081 Beograd – Zemun, Nemanjina 6, P.O.Box 127 Yugoslavia

Sombor cheese is a special type of domestic cheese with characteristic texture. The cheese ripening is taking place in a special Mould made of wood and very often the lower part of the Curd is more Acid than it is normaly desired. Some modifications in technological process were made in order to prevant this undesired characteristic. The cheese manufacture has been done in two variants (modification and control), from pasteurised milk with the addition 15% and 20% wather. The modified cheese production was consisted of washing the curd, immersion the cheese in the water at 35 C, and some changes in cheese ripening (18 C, 12 C and 85% R H). The result was not so intensive pH decrease. The data concerning NPN quantity show that the most intensive, increase was in first ten days and the lowest between 20–30 days.

The cheese manufactured with applied modification in tehnology had the shorter piriod of ripening, finer Body, Texture and better flavor and taste. It alsow had more free aminoacids then controle cheese.

SELECTED LACTIC CULTURES IMPROVE WHITE CHEESE HYGIENIC QUALITY

F.J. Chavarri, J.A. Nuñez, M. De Paz and M. Nuñez
Departamento de Bioquímica y Microbiología, Instituto Nacional de Investigaciones Agrarias, Apartado 8111, Madrid 28040, Spain.

Burgos cheese is the main white cheese variety manufactured in Spain. Although made from pasteurized milk no lactic cultures are added to the milk, and cheeses are not salted or are brine-salted for only 10-30 min. High moisture level and pH value and low NaCl content enhance growth of pathogenic and undesirable microorganisms. In the present work the influence of lactic culture inoculation on growth and survival of *Escherichia coli* in Burgos cheese was investigated, in order to improve its hygienic quality.
Pasteurized milk was inoculated in all cases with *E. coli* INIA 846 (100 cfu/ml). Two lactic cultures were used: culture N (*Streptococcus lactis* INIA 12) and culture B (*Str. lactis* INIA 12 + *Leuconostoc mesenteroides* subsp. *cremoris* NCDO 543 + *Leuc. mesenteroides* subsp. *dextranicum* INIA 141). Large (1.5 kg) and small (0.5 kg) cheeses were manufactured from all vats. Cheeses were stored at 0 C or 4 C during the first 24 h and at 4 C, 8 C or 12 C for seven additional days.

Storage temperature during the first 24 h and cheese size influenced significantly ($P<0.001$) growth of *E. coli* in Burgos cheese: mean log counts after 24 h were 2.60 for large cheeses and 1.85 for small cheeses if stored at 0 C, whereas the respective levels were 4.34 and 3.41 if stored at 4 C (each log count represents the average of 14 cheeses)

Analysis of variance on log counts of *E. coli* after 48, 72 and 96 h of storage detected a significant ($P<0.001$) effect of composition and level of starter, of cheese size and of temperature and time of storage.

After 8 days cheeses stored at 4 C or 8 C had acceptable pH values (over 5.25), but growth of *E. coli* had been restrained only in cheeses manufactured with lactic cultures. In control cheese, made without starter, log *E. coli* counts were in the range 4.88-6.36. In cheeses stored at 12 C and made with lactic cultures the pH was too low to be accepted by Burgos cheese consumers.

"CHYMOTEST-UGLICH" - THE NEW STANDARDIZATION SYSTEM FOR RENNET REFERENCE BATCHES

B.A. Surkov, V.A. Krayushkin
All-Union Research Institute of Butter and Cheesemaking, 152620, Uglich, Krasnoarmeyskii boulevard, 19, USSR

In 1984 the USSR Ministry for Meat and Milk Industry adopted a new automated system "Chymotest-Uglich" for the standardization of rennet reference batches. Instead of milk-clotting time determination apt to subjective errors the new procedure is able to measure mass content of chymosin and bovine pepsin on the basis kinetic relationships found for the turbidimetrically followed action of rennet on diluted milk.

TUESDAY - POSTER 40

THEORETICAL ASPECTS OF ARTIFICIAL COVERINGS' USAGE FOR CHEESE RIPENING

G.G. Shiller
All-Union Research Institute of Butter and Cheesemaking, 152620, Uglich, Krasnoarmeyskii boulevard, 19, USSR

Systematic analysis of interacting in complex "cheese-ambient space". Classification of objects in such complex. Role of water activity in inactivation of microbial aerosol and surface microbiocenosis of hard cheeses. Grafic and analytical models of inbitory processes. Functions and properties of artificial coverings for cheeses. Polycriterial system for grading of such coverings. Technological aspects of usage of latex, wax, film and combined coverings.

STRUCTURE-MECHANICAL CHARACTERISTICS OF RENNET CHEESES WHEN USING THE PROCESS OF MECHANICAL-IMPULSE-TREATMENT OF MILK

V.P. Golovkov, R.V. Ginoyan
All-Union Research Institute of Butter and Cheesemaking,
152620, Uglich, Krasnoarmeiskii Boulevard, 19, USSR

Mechanical-impulse-treatment of milk results in conformation changes of micellar structure of proteins and in the increase of the extend of dispersion of fat phase stipulating the shortening the time of curd structure formation by milk-clotting preparations by 25-35 per cent. The deep proteolysis and lipolysis of curd ensures the change of structure-mechanical characteristics of rennet cheese allowing 15-20 per cent viscosity decrease of the experimental cheeses as compared to the cheeses produced by traditional technology.

TUESDAY – POSTER 42

ANTAGONISTIC EFFECTS OF LACTIC ACID BACTERIA ON ENTEROBACTERIA

A.V. Gudkov, G.D. Perfiliev, N.P. Sorokina
All-Union Research Institute of Butter and Cheesemaking,
152620, Uglich, Krasnoarmeiskii Boulevard, 19, USSR

30 (88%) strains of Leuconostoc cremoris, 30 (51%) Lactobacillus plantarum, 6 (60%) L. casei, 8 (100%) L. bulgaricus, 21 (22%) Streptococcus lactis spp. diacetylactis, 15 (10%) S. lactis, 9 (13%) S. cremoris inhibited the growth of Enterobacteria in agar medium. Streptococci inhibited the growth of test-cultures in milk in log phase, lactobacilli and leuconostocci increased the death of Enterobacteria.

STARTERS: AS A MEANS OF CONTROLLING CONTAMINATING ORGANISMS

A.V. Gudkov
All-Union Research Institute of Butter and Cheesemaking,
152629, Uglich, Krasnoarmeiskii Boulevard, 19, USSR

Besides acids some species and strains of lactic acid bacteria produce the other antimicrobial substances. Lactic streptococci and lactobacillii strains producing the inhibitors against Enterobacteria and Clostridium were used for the manufacture of semi-hard cheeses from milk seeded up to 100 cfu/ml of M.coli or C. tyrobutyricum. These cheeses and cheeses made from unseeded milk were of the same quality.

TUESDAY - POSTER 44

PLANNING THE DAIRY ENTERPRISE TO MEET THE CHALLENGES OF THE MARKET

Rudolf Hilker
Meiereizentrale Nordmark e.G., Waidmannstrasse 10,
2000 Hamburg 50, Deutschland

Aufgrund der Forschungsergebnisse der Natur-und Ingenieurwissenschaften sowie deren Umsetzung in neue Technologien erhöht sich der Investitionsbedarf in der Molkereiindustrie stetig. Um das Risiko möglichst klein zu halten, bedürfen Investitionsentscheidungen stärker denn je einer Absicherung zum Markt. Im Referat wird auf Veränderungen, ausgelöst durch Bevölkerungsentwicklung, Ernährungs-und Einkaufsgewohnheiten, eingegangen.

NEW POSSIBILITIES IN ICE CREAM PRODUCTION BY INCORPORATING SOYA ISOLATE

M. CARIC, S. MILANOVIC, LJ KULIC, D. GAVARIC

Faculty of Technology, V. Vlahovica 2,
21000 Novi Sad, Yugoslavia

The production of ice cream mixtures, enriched with soya protein isolate, with flavouring agents : vanilla, rum punch and chocolate, was carried out. The aim was to increase the frozen desserts assortment by introducing new products of high nutritive value on to the Yugoslav market. The mixtures produced were spray dried and their quality determined.

Ice cream mixtures were composed of the following dairy components : pasteurized milk, sweet pasteurized cream, skim milk powder and whey powder; and non-dairy components : soya protein isolate, sucrose, commercial stabilizer/ emulsifier and flavourings : vanilla (0.2%), rum punch (0.2%) chocolate (2.0%) in permitted amounts. The investigated soya protein isolate was produced by Purina Protein Europe (PP-810 - min 90% proteins).

Ice cream mixtures were spray dried in semi-industrial plant according to the following procedure :
Mix preparation (min 30% TS; min 8% milk fat) → mixing and dissolving (50°C) → filtration → homogenization (65-70°C; $9,8 \times 10^6$-29.4×10^6 Pa) → Pasteurization (85°C; (min) → spray drying (t_{inlet} 220°C; t_{outlet} 90°C) → packaging.

TABLE 1. Ice cream mix composition (neutral samples)

Component %	Sample 1	Sample 2
Milk	71.5	71.5
Cream	10.07	10.07
Skim milk powder	2.28	1.14
Whey powder	-	1.14
Soya protein isolate PP-810	1.14	1.14
Sucrose	14.5	14.5
Stabilizer/ emulsifier	0.5	0.5

Note : Sample 1 contains skim milk powder and PP-810 (2:1);
Sample 2 contains skim milk powder, whey powder and PP-810 (1:1:1)
The whole quantity of mentioned three components being the same in both samples.

TABLE 2. Chemical composition of dairy components used in ice cream production

Component	TS%	Fat %	Proteins %	Minerals%
Milk	11.13	3.3	-	-
Cream	56.90	56.5	-	-
Skim milk powder	96.95	7.7	33.10	8.30
Whey powder	96.71	3.55	16.68	7.59

TABLE 3. Chemical composition of ice cream mix, enriched with soya protein isolate PP-810

Sample	TS%	Fat %	Proteins %	Minerals %	pH
1	36.01	9.27	5.85	0.82	6.62
2	34.37	9.37	5.12	0.82	6.62
Control (without PP-810)	33.60	9.50	4.03	0.82	6.40

Total solids content in two experimental neutral samples is above 30% and total fat is above 8%. What are mineral values for ice cream mixtures produced in Yugoslavia, both samples contain more proteins, compared to the control, due to soya protein isolate addition.
Both samples with; vanilla, rum punch or chocolate flavouring, were organoleptically evaluated and deserved high ratings.

TABLE 4. Chemical composition of dry ice cream mix, enriched with soya protein isolate PP-810

Sample	TS%	Fat %	Proteins %	Minerals %
1	97.56	33.8	11.56	2.27
2	98.77	34.0	11.68	2.25

Spray drying of ice cream mixtures, enriched with soya protein isolate resulted in dry ice cream mixes of excellent quality, low water content in both samples (2.44% and 1.23% respectively) enables long storage even at room temperature. Dry mixes were in the form of fine powder with pleasant aroma and taste.

Chemical and organoleptic analysis demonstrated the high quality of the newly-developed ice cream based dairy products enriched with soya protein isolate.

Modified dry ice cream mixtures contain besides animal proteins, pure plant proteins, which contribute to physico-chemical (functionality-consistency) and nutritional quality of the product, being also economically justified.

ISOLATION AND CHARACTERIZATION OF A WIDE HOST RANGE PHAGE FROM A MESOPHILIC STARTER CULTURE

A. von Wright, T. Saarinen, and S. Sivelä
Valio Finnish Co-operative Dairies' Association
Research and Development Department
P.O. Box 176
SF-00181 Helsinki
Finland

A mesophilic cheese starter from a Finnish dairy plant was divided into 32 streptococcal strains on the basis of their different plasmid complements. After a suspected phage outbreak each strain was exposed to a whey sample, and 23 different phage isolates, according to the host specificity, were obtained. The DNA restriction patterns, protein composition, and general morphology of these isolates were compared.

In EcoRI digestion a characteristic restriction pattern with four DNA-fragments (14 000, 6 200, 3 150, and 2 500 base pairs) were obtained from the majority (16) of the isolates. Among the rest of the isolates there was some variation in the sizes and number of the DNA-fragments. There were, however, enough similarities between the restriction patterns to conclude that all 23 phage isolates were closely related.

In SDS-polyacrylamide gel electroforesis the phage variants, despite the differences in DNA-restriction patterns, had all five major protein bands with molecular weights ranging from 25 000 to more than 70 000.

In electron microscopy the phage variants were morphologically identical, each having an oval 55 x 40 nm head and a 90 nm tail.

It was concluded that the phage isolates were all variants of a single phage species with a wide host range (*Streptococcus lactis*, *Str. lactis* subsp. *diacetylactis*, *Str. cremoris*). The differences in host specificities among the variants most probably reflect different restriction-modification systems or other defence mechanisms operating in the host bacteria. It is evident that host specificity alone is a very unreliable criteria in studying the taxonomy of phages in dairy environment.

SPECIATION OF SELENIUM IN COW'S MILK

A. Deschuytere, K. Vermeylen, O. Van Schoor, H. Deelstra and M. Naudts*

University of Antwerp, Universiteisplein 1, 2610 Wilrijk, Belgium
*Rijkszuivelstation, Brusselsesteenweg 370, 9230 Melle, Belgium

INTRODUCTION

Since 1973 selenium is generally accepted to be an essential trace element for man. The favourable influence of Se in the prevention of Keshan disease, its rule in mucoviscidose and possibly other diseases, led to the recognition of the importance of the presence of this trace element in infant formulas. However not only the total amount of Se in the food is of interest, but also the bioavailability must be considered.
Since the bioavailability of an element is usually dependent upon its chemical form, in this work the major Se compounds in aqueous solution were separated. The major protein fractions of cow's milk were analyzed for their Se concentration. After enzymic proteolysis of these fractions, the different Se compounds will be separated.

TABLE I. SELENIUM CONCENTRATIONS OF THE DIFFERENT PROTEIN FRACTIONS OF COW'S MILK

	%	ppb Se	absolute Se conc. in each fraction (ng)
WHOLE MILK	100	11-14	1100-1400
SKIMMED MILK	96.5	9-13	868-1255
FAT	3.5		
PROTEINS	3.5-4.0		
• Casein	2.4-2.8	140-160	336-448
α-casein	1.5-1.9	93-99	140-188
β-casein	0.9-1.1	144-160	130-176
κ-casein	0.3-0.4	426-432	128-173
γ-casein	0.1-0.2		
• Whey proteins	0.5-0.7	610-720	305-504
α-lactalbumin	0.1-0.15	200-350	20-52
β-lactoglobulin	0.2-0.4	800-808	160-320
immunoglobulins	0.06-0.10		
proteose-peptone	0.06-0.18		

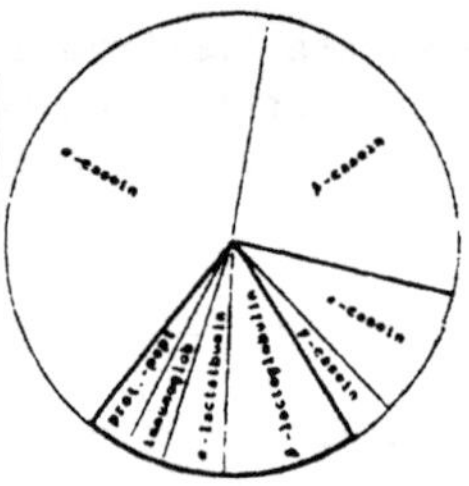

FIGURE I a). ABUNDANCE (%) OF THE DIFFERENT PROTEIN FRACTIONS OF COW'S MILK

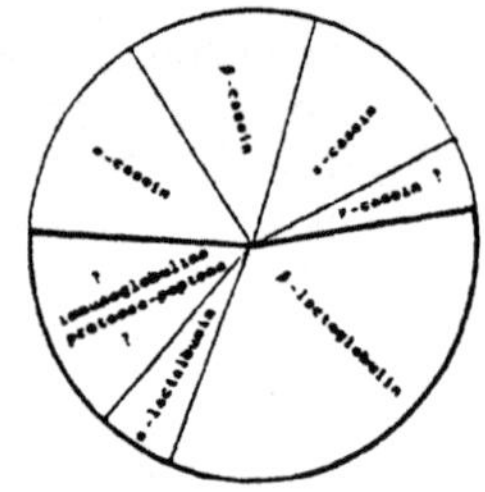

FIGURE I b). THE RELATIVE DISTRIBUTION OF SELENIUM OVER THE DIFFERENT PROTEIN FRACTIONS OF COW'S MILK.

TABLE II. SELENIUM CONTENT OF HUMAN MILK AND COW'S MILK AT DIFFERENT LACTATION STAGES.

	mean (SD) ng/ml of cow's milk	mean (SD) ng/ml of human milk
COLOSTRUM MILK (1-4 days post partum)	23.1 (5.8)	14.8 (4.9)
TRANSITIONAL MILK (6-10 days post partum)	15.7 (5.3)	12.3 (4.4)
MATURE MILK (10 days - 12th month post partum)	11.8 (3.4)	9.65 (2.7)

TABLE III. SEPARATION METHOD FOR 4 DIFFERENT SELENIUM COMPOUNDS IN AQUEOUS SOLUTION.

SYSTEM	: BAKER-10 spe
COLUMN	: DOWEX-1 ionexchange column
PRETREATMENT	: CLEANING with • 2 X 1 ml NH_4OH / H_2O (1:1) REGENERATION with • 2 X 1 ml H_2O • 2 X 1 ml 1N HCl • 2 X 1 ml 0.1N HCl • 2 X 1 ml 0.1N Tris HCl buffer (pH = 5.52 +/- 0.01)
SAMPLE	: INTRODUCTION of 1 ml of sample onto the column (contact time of a few minutes)
ELUTION	: FIRST with 4 X 1 ml 0.1N Tris HCl buffer (pH = 5.52 +/- 0.01) ⟶ Se-METHIONINE AND Se-CYSTINE elute FOLLOWED BY 5 X 1 ml 0.1N HCl ⟶ SeO_3^{2-} elutes SeO_4^{2-} remains on the column

CONCLUSION

In milk, most of the Se is associated with the proteins. Cow's milk contains more proteins than human milk, however the Se concentration of transitional and mature cow's milk does not differ with that of human milk. Possibly human milk contains only the Se richer protein fractions.
The Se amino acids, selenite and selenate in small volumina of an aqueous solution were separated. The method proposed is a fast, efficient and cheap method which allows parallel elution of several samples at the same time. However by this method selenocystine and selenomethionine eluted together and further research for their separation is necessary. The bioavailability of Se in both cow's milk and human milk protein fractions, using this separation method, will be looked for in future research.

Oral Intake of Glucose plus Galactose and Erythrocyte Galactose-1-Phosphate in Man. - A Nutritional Evaluation of Hydrolyzed Lactose

C.A. Barth and N. Kopra
Bundesanstalt für Milchforschung, D-2300 Kiel 14
Postfach 6069, Federal Republic Germany

This study deals with the metabolic effects of hydrolyzed lactose:

5 healthy adult volunteers consumed after an overnight fast a glucose-galactose mixture equivalent to 61.4 g of lactose (or 125 g of dried skim milk powder with hydrolyzed lactose). The postprandial rise of erythrocyte galactose-1-phosphate (gal-1-P) never exceeded 22.3 µmol per liter packed red blood cells (Fig. 1). This amounts to no more than 22 % of the levels known from galactosemic children to be safe, concerning ocular, neural or hepatic damage. We conclude that the consumption of the hydrolyzed lactose does not cause a risk for consumer's health as judged from this galactose metabolite. A considerably higher risk, however, may accompany the consumption of galactose alone which causes around 17-fold higher plasma galactose levels and around 8-fold higher erythrocyte gal-1-P concentrations for more extended time periods.

Caution is advisable if technologies are used which lead to a mixture of galactose plus fructose (instead of glucose) (Williams, C.A. et al. (1983) Metabolism 32, 250-256).

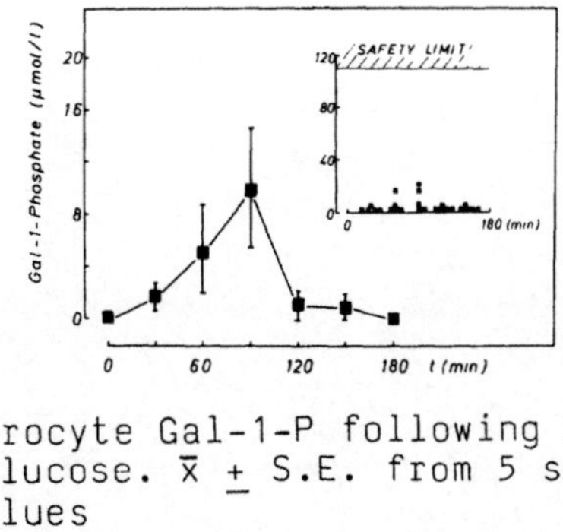

Fig. 1: Erythrocyte Gal-1-P following 32.5 g galactose plus 32.5 g glucose. $\bar{x} \pm$ S.E. from 5 subjects. Insert: individual values

Microbiological assay of antibiotics in Egyptian Dairy products.

S.A.Abou-Donia, M.M.Rida, S.E.Rakshy
Department of Agricultural Industries,
Faculty of Agriculture, University of
Alexandria, Egypt.

Sensitive strains of Streptococcus thermophilus were used for assay of antibiotics in Egyptian dairy products. It was found that the yeast extract lowered the sensitivity of the above strains to penicillin. It was found that 3.1% of milk samples in Alexandria were positive to antibiotics Both of examined pasteurized, dried, evaporated, sweetened condensed, sterilized and UHT milk were found to be negative. The high sensitivity of S.thermophilus starter to NaCl gave false results. The penicillin remained detected for three days in the milk after being ingected cows suffering from mastitis. Adding of penicillin ranged 0.05 -0.30 IU/ml milk caused an apparent effect. The cells showed tendency to be longer and thicker than unexposed cells. Penicillin gave stability in both pasteurized and boiled milk. When Bacillus stearothermophilus used it was found that the later organism more sensitive than the former for the assay of antibiotics.

EFFECT OF SOME STABILIZER MIXTURES ON THE PHYSICAL AND ORGANOLEPTIC QUALITIES OF KAHRAMANMARAS TYPE ICE CREAM

O.C. Tekinsen, *A. Karacabey
Selcuk Üniversitesi, Veteriner Fakültesi, Konya, Turkey
*Ankara Üniversitesi, Ziraat Fakültesi, Ankara, Turkey

Of the samples overrun increased up to 35,2% while melting resistance decreased as the level of salep increased gradually from 50 to 100% in the stabilizer mixture. Organoleptic properties of the all samples differed significantly depending on the mix composition and the level of salep in the stabilizer mixture. The highest overall score was observed in the sample manufactured from the modified mix with the stabilizer mixture containing 75% salep.

NITROGEN DISTRIBUTION , MOLECULAR WEIGHT AND SIZE OF CASEIN MICELLES IN AWASSI SHEEP'S MILK

R.M. SALEEM , W. A. MAHMOUD AND Y. A. YOUNIS

Casein content formed 80% of total protein while the albumin , globulin and proteose - pepton were 10.5 , 4.8 and 4.7% respectively . The non - protein nitrogen content formed 4.4% of the total milk nitrogen .

The average molecular weight and size of calcium casenate phosphate particales were 202.18 million and 81.34 mu. respectively . More than 50% of the samples molecular weight ranged from 150 - 250 million , while about 90% of samples had particle size ranged between 70 - 100 mu.

THE COMPOSITION OF BUFFALO'S MILK IN THE MOSUL AREA

B.K. DALALY, L. ABD-EL-MOTTALEB, A. EL-SHAZLY & R. ABDALLAH

Department of Food Technology,
College of Agriculture and Forestry, Mosul University,
Hammam Al-Alil, Mosul, IRAQ.

Fourty nine individual buffaloe's milk samples were collected frum Mosul area. The samples were analysed and the range for pH 6.40-7.00,% acidity 0.12-0.27, fat 5.20-11.oo, S.N.F. 8.86-11.70,total nitrogen 0.407-0.798, lactose 4.63-5.80 and ash 0.61-1.28.

EFFECT OF STAGE OF LACTATION AND SEASONAL VARIATION ON THE CHEMICAL COMPOSITION AND PROPERTIES OF CAMEL'S MILK

A. Wahba, F. El-Abbassy, I. Ismail, S.I. El-Agamy
Department of Agricultural Industries, Faculty of Agriculture, University of Alexandria, Egypt

Camel's milk samples were collected from different rural areas round Alexandria. The TS, SNF, lactose and all nitrogenous components were always considerably higher in winter milk than in summer milk. Chloride and titratable acidity were higher in summer and during the late period of lactation.

WEDNESDAY – POSTER 8

COMPARATIVE STUDIES ON THE MILK CONTITUTES OF HUMAN, CAMEL AND GOAT MILK

A. Wahba
Dairy Department, Faculty of Agriculture,
University of Alexandria, Alexandria, Egypt

Camel and goat milk are common to be used for infant feeding in most of the rural areas through the Nile valley. Camel milk is nearer in its chemical and physical properties to human milk than goat milk. The B-lactoglobulin and proteose -peptone nitrogen contents were equal values in human and camel milk.

WHY IS IT DIFFICULT TO MAKE CHEESE FROM CAMEL MILK?

Märtha Larsson-Raźnikiewicz, * Muctar Ali Mohamed
Dept Chem & Mol Biol; Swedish Univ of Agric Sci,
P.O. Box 7015, S-750 07 Uppsala, Sweden
* Faculty of Industrial Chem, Somali Nat Univ,
P.O. Box 1081, Mogadishu, Somalia

The far-away aim is to provide new information on the nutritive qualities and the properties influencing the technological properties of camel milk; to make camel milk and milk products available to a wider range of consumers in arid and semi-arid zones in tropical and subtropical countries, especially against a background of substantial seasonal overproduction. Attempts to make cheese from camel milk have until now been associated with difficulties (see Yagil, R. 1982 in FAO Animal Production and Health Paper 26, chap 4).

Calf rennet was used to test for possible differences in the curd formation of cattle and camel milk. The small amount of rennet used to curdle cattle milk appeared to have no effect on camel milk. When the enzyme quantity was increased about 50-fold a precipitate of the milk proteins was obtained - no gelatinous curd as with cattle milk. The reason to slowness of the rennet reaction and the low water-binding capacity of the camel "curd" is of special interest. The composition and properties of the casein (cheese protein) fractions as well as the mineral composition certainly are important for the technological behaviour of milk. The same can be said about the general composition.

Full casein from fresh camel milk was analysed for the various casein components. Identification by gel electrophoresis and amino acid and phosphorous analyses showed that as in cattle milk the four principal caseins α_{s1}-, α_{s2}-, β- and κ-casein exist also in camel milk. The cow and camel caseins are similar but for example charge differences occur, making the gel electrophoresis patterns look different, for example, α_{s2}-casein moves in front of α_{s1}-casein in electrophoresis at alkaline conditions. As κ-casein makes the caseins in milk form a homogeneous solution and curd formation by rennet action is primarily due to an attack on κ-casein especial attention must be paid to the κ-casein status in camel milk (cf. M L-R, MAM, Swedish J. agric. Res. (1986) 16, 13-18). In many camel milk samples κ-casein is difficult to identify with conventional techniques.

MANUFACTURE OF KHOA, AN INDIAN MILK PRODUCT, BY REVERSE OSMOSIS

DHARAM PAL* and MUNIR CHERYAN**
*National Dairy Research Institute, Karnal, India.
**University of Illinois, Department of Food Science,
382D A.E.S.Building, Urbana, Illinois 61801, USA

INTRODUCTION

The use of synthetic membranes in the food and dairy industry is rapidly increasing around the world [1]. To date, the major applications on a world-wide basis is the use of ultrafiltration for fractionating cheese whey or for pre-concentrating milk prior to cheesemaking. In India, however, owing to the nature and demand of indigenious dairy products, reverse osmosis, which is essentially a dewatering operation, should have wider application.

Khoa is an important indigenious milk product. It has been estimated that over 2500 million kg of milk (as much as 7% of India's total milk production)is converted to khoa per year in India [2]. It is presently manufactured on a small scale by continuous boiling of milk until a desirable solids concentration (65-70% total solids) is reached. Several attempts have been made to develop new methods for commercial production of khoa, including the use of scraped-surface kettles or heat exchangers, evaporation for partial moisture removal [3] and use of dehydrated milk [4]. The general conclusions were that using pre-concentrated milk above 31% total solids resulted in poor quality khoa. The superior energy efficiency of reverse osmosis during initial stages of milk concentration could be used to advantage in developing an economical low-energy continuous khoa-manufacturing process. The objectives of this study were (1) to optimize the processing of milk by reverse osmosis, (2) to manufacture khoa from RO-concentrated milk and evaluate product quality, and (3) to evaluate the energy aspects of the process.

MATERIALS AND METHODS

Cow's milk was obtained from the University of Illinois dairy herd. Reverse osmosis was used to pre-concentrate the milk to either approximately half its initial volume (2X) or to 40% of its initial volume (2.5X). The concentrated milk was then converted into khoa in an open stainless steel kettle using a teflon-coated stirrer. Khoa made by the traditional open-pan method served as the control. The following variations in the basic process were studied:

Process A: Raw milk, with no further treatment, was processed in the reverse osmosis module prior to open-pan boiling.
Process B: Milk was pasteurized at 63C for 30 min before reverse osmosis. The process was otherwise identical to Process A.
Process C: Milk was pasteurized and separated in a cream separator. The skim milk was concentrated by RO. The RO retentate was then combined back with the cream before converting it to khoa.
Process D: Milk was pasteurized and homogenized in a Manton-Gaulin homogenizer at a total pressure of 2000 psig (135 bar). The homogenized milk was concentrated by reverse osmosis.

Reverse osmosis was done on a spiral-wound pilot-scale module (192-SEPA 97, with 1.4 sq.m. effective membrane area) manufactured by Osmonics Inc, Minnetonka, Minnesota, USA. A cellulose acetate membrane, rated at 97% NaCl rejection, was used. Unless otherwise mentioned, RO was done at 30C and pH 6.8. For these trials, the RO system was operated in the batch mode. During the fouling and optimization studies, both retentate and permeate were recycled to the feed tank. During manufacture of the pre-khoa, the permeate was discarded and the milk allowed to concentrate.

RESULTS AND DISCUSSION

A linear relationship between flux of pure water and transmembrane pressure was obtained. The average pure water permeability coefficient for this membrane is 1.45 liters/sq.m./hr/bar. Milk, however, showed a significant departure from linearity and its flux was much lower than that of water (Figure 1). There are two possible reasons for this: (i) No permeation will occur until the applied pressure exceeds the osmotic pressure of the retained solids. Due to the presence of the solids in milk, especially the dissolved salts and

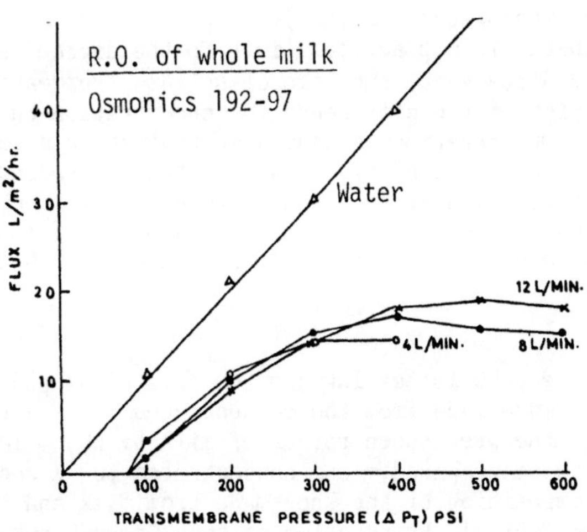

Figure 1. Effect of transmembrane pressure and flow rate on flux.

lactose, the osmotic pressure of milk is about 90-100 psi (6.1-6.8 bars). This explains why there is essentially no permeation until the transmembrane pressure is about 90 psig (Figure 1) and why flux of milk will always be less than that of water.

(ii) In addition to osmotic pressure, another resistance must be overcome due to concentration polarization and the associated boundary layer. Improvements in flux could only be obtained by reducing solute concentration at the membrane surface, e.g., by using high shear at the surface or by increasing the turbulence in the module. This explains why higher flow rates improved the flux with milk (Figure 1). The fat and protein contribute little to the osmotic pressure compared to the salts and lactose, but will be the major reason for the increased hydrodynamic resistance. Greater polarization will occur at higher pressures, and thus the effect of flow rate will be more noticeable at higher pressures.

The flux of skimmilk was usually higher at equivalent pressures and flow rates than whole milk [not shown here: see ref.6]. Homogenization improved the flux at the lower flow rates, but was not beneficial at the higher flow rates. It appears that the optimum pressure is 400 psig (27.3 bars) and optimum flow rate with this module is 12 L/min, which is the highest flow rate recommended by the manufacturer.

Our data generally followed the linear relationship between flux and the log of the bulk solids concentration expected from the film theory:

$$J = k \ln (Cg/Cb)$$

where k is a mass transfer coefficient, Cb is the bulk solids concentration and Cg is the "gel" concentration at the membrane surface. By extrapolation, we obtained a Cg value of approximately 36-38% total solids for the RO of milk under these conditions. The mean k value was 13.5 liters/sq.m./hr (LMH).

However, there is a practical limit to the degree of concentration which is dictated by the loss of solids. Permeation of the solids is a function of the difference in concentration on either side of the membrane. Our experiments indicated that above a concentration factor of 3X, the "leakage" of ash and lactose increased significantly [6]. However, even at the highest retentate concentration, the overall rejection of the solids was more than 95%. If losses of the solids are to be minimized, the concentration factor should be kept below 3X.

Product Quality

In order to keep RO losses low and the flux high, and the sensory properties of the khoa made from the concentrates satisfactory, it was decided to limit the pre-concentration of milk by RO to less than 3X. Two concentrates were prepared, one concentrated to 2X and the other to 2.5X. The composition of the khoa made from milk and its concentrates is shown in Table 1. Preparation of khoa from RO retentates took less than half the time of the control khoa made by the traditional method. In general, khoa from RO concentrates had higher moisture contents than the controls. The fat content of all khoa, including the control, was usually less than the minimum required 20%, but fat content can be adjusted during the process, either before or after RO.

A significant difference was observed in the fat and ash concentrations between control and RO-khoa. The ash was always lower in the RO-khoa. This could be due to the passage of some ionic species through the membrane during concentration. At a concentration factor of 2-2.5X, rejection of ash was typically about 70% and thus the lower ash contents of the RO-khoa is to be expected. Lactose loss was much less since the rejection of lactose was in excess of 97%. Protein was measured as Kjeldahl nitrogen x 6.38. Some nonprotein nitrogen may pass through the membrane, but protein contents of RO-khoa were comparable to the control. Fat rejection is essentially 100% with RO membranes. Since there was a loss of other components, the fat content of RO-khoa is expected to be slightly higher.

Table 2 shows sensory properties of RO-khoa. There were essentially no differences in the Hunter colour values (L,a,b) between the RO-khoa prepared by the 4 processes or whether it was prepared from a 2X or 2.5X retentate. The RO-khoa showed less of a greenish tinge than control khoa, but the differences were not significant. There was a noticeable yellowish-orange colour to all khoa samples, due to the presence of carotene in the cow's milk.

There were no differences in other sensory properties between 2X and 2.5X RO-khoa samples. The RO-khoa samples did not show any large differences in flavour and body from their respective controls. Since Process A used raw milk, both the control and the RO khoa were rancid. This is due to the hydrolysis of fat by lipase which was not inactivated before concentration. RO-khoa tended to lack graininess, but

Table 1. Composition of khoa (as-is basis)

Process	Concentration factor	Total solids(%)	Fat (%)	Protein (%)	Ash (%)	Lactose (%)
A	Control	69.9	17.0	19.9	4.4	28.5
	2X	63.2	16.1	18.1	3.7	25.3
	2.5X	60.5	15.7	17.4	3.4	24.1
B	Control	67.7	19.6	19.3	4.0	24.9
	2X	61.4	18.6	17.7	3.2	22.0
	2.5X	60.3	18.3	17.5	3.0	21.5
C	Control	66.7	19.3	18.8	4.3	23.8
	2X	66.4	20.6	18.8	3.6	23.3
	2.5X	65.9	20.5	18.7	3.5	23.2
D	Control	62.5	17.8	17.4	3.9	23.5
	2X	61.3	18.1	17.1	3.3	22.8
	2.5X	60.6	17.9	17.0	3.2	22.5

Table 2. Sensory properties of RO-khoa

COLOUR BY HUNTER COLORIMETER (average of all trials)

	L	a	b
Control khoa	69.3	- 4.1	19.5
RO-khoa (2X)	67.0	- 1.5	16.7
RO-khoa (2.5X)	67.8	- 3.1	15.7

FLAVOUR AND TEXTURE (average ratings for all samples)

Control khoa*	: Normal, granular, typical flavour
RO-khoa [Process A]	: Rancid, free fat, lacks grains
RO-khoa [Process B]	: Normal, free fat, no grains
RO-khoa [Process C]	: Cooked, slight free fat, no grains
RO-khoa [Process D]	: Normal, no free fat, no grains

* Except for Process A control, which had a rancid flavour

this may not be an undesirable quality; rather, in preparation of products such as burfi, this will help in producing a homogeneous and smooth product. It is possible the graininess could be controlled during the final finishing/heating stages. Except when homogenized, the RO-khoa showed slightly higher free fat which was evident from deposition on the sides of the container.

Energy Considerations

The following assumptions have been made with all process calculations (figures have been rounded off in some cases):

(1) Raw milk is available at 25C at a solids concentration of 15%.

(2) Pasteurization of milk,primarily to inactivate the lipase, is done in a high-temperature, short time (HTST) system at 75C with a regeneration efficiency of 80%.

(3) Milk is processed at a rate of 20,000 kg milk per day producing 4615 kg of khoa at 65% total solids. The plant operates for 20 hours per day, with about 4 hours per day for clean-up and maintenance.

(4) Preconcentration is done to 31% total solids before finishing in the atmospheric boiler. This implies that 10,323 kg of water is removed by reverse osmosis, leaving only an additional 5062 kg of moisture to be removed in the boiler per day.

(5) It was assumed that steam turbines were used to generate electricity for the pumps, and that the conversion of steam energy into electrical pump energy was 50%.

(6) The cost of steam generated in India is Rs.0.30 per kg of steam [7].

(7) Average flux for a batch operation for concentrating milk up to 31% total solids was 8 LMH.

The methods suggested by Cheryan [5] were used to calculate energy consumption for the RO processes. We considered two modes of operation: (a) batch operations, using either a single pump with total recycle of retentate or a dual pump system with internal recycle of retentate, and (b) continuous feed and bleed operations, using either one stage or 3 stages. A multi-stage recycle system removes only a part of the permeate in each stage. Hence, all stages except the last one will be operating at much higher flux levels. We assumed the first stage is operating at 20% solids and the second stage at 26% solids with corresponding fluxes of 9 and 6.2 LMH respectively. The third stage will be operating at the final required solids level of 31%, where the flux is 2.5 LMH.

Table 3 is a comparison of all of the processes. It is obvious that energy consumption will be markedly reduced using membranes to partially pre-concentrate the milk. On average, it can be assumed that energy consumption of a RO operation for khoa will require about 20 kcal/kg of milk, 75% of which is for the pasteurization alone. An additional 136 kcal/kg will be required for the open-pan boiling to 65% solids.

The feasibility of installing such systems is best considered on two criteria: the payback period, which is the time required to recoup the initial investment in the plant on the basis of energy savings, and the potential profit margin, which is the difference between the selling price of the product and the production costs. The latter will include operating costs, cost of milk and plant depreciation.

As far as the payback period is concerned, the savings in energy from a khoa production plant processing 20,000 kg/day of milk that uses RO instead of only open-pan boiling will be 326 (i.e.,346-20) kcal/kg of milk, which translates into about 6,520,000 kcal per day. This means a savings of more than 12,000 kg/day of steam. At Rs. 0.30 ($0.024) per kg, this is a savings of Rs 3600 per day in steam costs alone. One manufacturer in India recently quoted a price of approx-

Table 3. Comparison of energy consumption and surface areas for pre-concentration of milk to 31% total solids at a rate of 1000 kg/hr

Process	Membrane Area (square meters)	Energy (kcal/kg milk)
Open-pan boiling*	--	345.9
Membrane Process:**		
Batch, single pump	65	47.9
Batch, dual pump	65	4.0
Continuous, one-stage	206	9.7
Continuous, three-stage	93	4.4

* Does not include energy for agitation or scraping.

**An additional 15 kcal/kg of milk will be needed for pasteurization in a HTST plate system of 80% regeneration

imately Rs. 300,000 ($25,000) for a 70 sq.m. spiral-wound RO system, which would imply a payback period of 3 months or less. Not included in these figures is the additional savings in the scraped-surface vessel used to finish the khoa, which can be reduced in size by 67% since only 1/3 the water has to be removed. Stabile [8] recently arrived at a similar conclusion that, in USA, the capital cost of a multiple-effect evaporator for partial removal of water from milk is the same as the capital cost of a reverse osmosis plant of the same capacity, but the energy consumption was much less for a 2-3X concentration of skimmilk.

CONCLUSIONS

The application of reverse osmosis for the manufacture of khoa appears to have great potential in India. The khoa prepared from RO-concentrated milk has a slightly higher moisture content, but the flavour and body does not appear to be affected, according to customary sensory evaluation criteria [9]. This should result in higher yields of the product. The potential savings in energy and overall economics appears to be extremely attractive.

REFERENCES

1. Cheryan, M. 1984. Membrane processing. In McGraw-Hill Yearbook of Science and Technology, 1985 edition, Parker,S.P.(Ed.), McGraw-Hill Book Co., New York. p.182.
2. Rajorhia, G.S. and Srinivasan, M.R. 1979. Technology of khoa--a review. Indian J. Dairy Sci., 32: 209.
3. Voghra, V.A. and Rajorhia, G.S. 1982. Utilization of pre-concentrated milk for khoa making . Asian J. Dairy Res., 1: 1.
4. Voghra, V.R. and Rajorhia, G.S. 1983. Utilization of dried milk for khoa making. Asian J.Dairy Res., 2: 113.
5. Cheryan, M. 1986. Ultrafiltration Handbook, Technomic Publishing Co., Lancaster, PA.
6. Pal D. 1985. Report on UNDP Fellowship at University of Illinois, NDRI, Karnal, India.
7. Sarkari,T.R. 1983. Heat recovery systems practical for industry, Chem.Age of India, 34(2): 91.
8. Stabile, R.L. 1983. Economics of reverse osmosis and multistage evaporation for concentration of milk from 8.8% to 45% total solids. J.Dairy Sci., 66: 1765.
9. Pal,D. and Gupta,S.K. 1985. Sensory evaluation of Indian milk products. Indian Dairyman, 37(10): 465.

ACKNOWLEDGEMENTS

This project was supported in part by the Illinois Agricultural Experiment Station and United Nations Development Program of the Food and Agricultural Organization (Project No. IND/78/020), which provided Dr. Dharam Pal with a Fellowship to conduct the research at the University of Illinois, Urbana.

IMPACT OF MILK MARKETING THROUGH DAIRY COOPERATIVES

R.L. KHERDE and R. SUBRAMANIAN
Head, Division of Dairy Extension
National Dairy Research Institute,
Karnal-132001 (Haryana) INDIA

The impact of Dairy Cooperatives on knowledge, adoption and production and productivity of animals was examined. The study revealed that the knowledge and adoption of Dairy Innovations was significantly higher in cooperatives areas. There was significant reduction in age at first calving, dry period, calving interval and mortality rate in buffaloes. Cooperatives generated more employment per farm per annum than control villages.

EFFICIENT MILK TRANSPORTATION SYSTEM

K.K. Kalra, Raj Vir Singh
Dairy Economics, Statistics and Management Division,
National Dairy Research Institute, Karnal 132001 (Haryana)
India

Efficient milk pick-up routes were developed with the help of vehicle scheduling model for transportation of milk from milk shed area to the plant. Monthly transportation route planning was done separately for milk assembling and milk chilling centres. The various conditions and restrictions typically found in India were considered for the routes system. On an average, the trucks and tankers taken together were travelling about 1014 km per day in the existing milk pick-up system. The efficient routes formulated reduced the daily travel distance to 954 km., saving through rationalised routes by about 6% (60 km. per day). This showed that the existing route system of the plant which was also most of the time in circular path was reasonably good and could further be improved by efficient routes planning.

IMPROVING DAIRY PLANT ECONOMY - SYSTEMS ANALYSIS

K.K. Kalra, Raj Vir Singh
Dairy Economics, Statistics and Management Division,
National Dairy Research Institute, Karnal 132001 (Haryana)
India

Vehicle scheduling, specially designed linear programming and interval programming models were used to optimise the three inter-dependent operations of procurement, processing and distribution of a dairy plant. The optimisation of processing and distribution along with efficient milk pick-up routes formulation decreased milk procurement cost by Rs 0.12 per 100 kg milk received at the dock of the main factory and the total variable cost of plant by Rs. 1.93. It further increased the gross revenue and the total contribution value (revenue over variable cost) by Rs. 8.52 and 10.45 respectively per 100 kg milk handled by the plant. The optimal results were examined for their sensitivity with respect to variation in restriction levels and products prices and also to establish the extend of advertisement expenditure to be made on each product for sales promotion.

MONGOLIAN TRADITIONAL MILK FOOD WURUM---A TYPE OF HARDEND AND CRISP CREAM LAYER CAKE

By JIN SHI-LIN

Inner Mongolia Light Industry Scientific Research Institute, 30 East Hairlar Road Huhhot
Inner Mongolia P.R. China

SUMMARY. WURUM is a type of Mongolian traditional milk food which is like hardend and crisp cream layer cake, with a semicircular shape, diameter is about 10 cm and 1 cm thickness, both sides surface covered with milk protein layer possess a lot of pockmarked face and sandwiched in between the two sides is the milk fat which is not exposed to light, thus it can prevented oxidation by the protein layer.

Technological process is introduced. The composition of the product is about 20% milk protein, 30% milk fat and with a moisture of about 10%.

Milk product are the principle foods in Monglian nationality, as importance as the meat. Many kinds milk products made from cow's milk. Sheep's milk, goat's milk, mare's milk and camel's milk. Ghee and hard cheese made from cow milk and goat milk can preserve a long time while many kinds of yoghurt and WURUM made from cow milk and goat milk only be preserved a short time. Sheep milk and camel milk are used to make tea with milk, mare milk is specialize to make koumiss. From mongolian experience you can use the mixed milk of cow's milk with sheep's milk but the mixture of cow's milk and camel's milk or goat's milk is not favourable.

WURUM in the traditional milk foods of Mongolia may be a kind of very nice Snack food of to day.

The process of WURUM

WURUM with a semicircular shape, diameter is about 10-15 cm and 1-1.5 cm thickness,both sides surface covered with milk protein layer pkossess a lot of pockmarked face and sandwiched in between the two sides is the milk fat which is not exposed to light, thus it can prevented oxidation by the protein layer. The composition of the product is about 20% milk protein, 30-40% milk fat and with a moisture of about 10%. Preserving period of time in room temperature is about one mouth, if vacuum packed with polyfoil will be preserve a long time.

The block diagram in Fig.1 summary the processing of WURUM.

```
| Raw milk |
    ↓ filteration purefication
| Heating in many pot with 15-20 cm diameter |
    ↓
| Bubbling and Foaming |
    ↓
| Cooling |
    ↓
| Defoaming |
    ↓
| Form a protein layer |
    ↓
| Set for creaming |
    ↓
| Form a cream layer |
    ↓
| Cooling |
    ↓
| Form a shell |
    ↓
| Drain off the skim milk |
    ↓
| Take up the shell and folding |
    ↓
| Cooling |
    ↓
| Packing with polyfoil |
```

Fig.1 Processing diagram of WURUM

Theoretical aspect about the process of WURUM

Bubbling and foaming and creaming are the key process of manufacture.

Bubbling and foaming

Foam in every respacts may be alike to o/w type emulsion, the density of foam is much low to which of milk, therefore a stronger buoyancy force to the bubble foam, the air bubbles in the foam can deform the bubbles each other from circular to polyhedral. All of the bubbles take a same size crowded together and packed closely layer upon layer. In the process of WURUM is quite the same.

Factors affect the foam stability

1. Milk protein especially casein micelle get in touch with air-milk interface, usually unfold its peptide chains and molecules interact eath other to make a two-dimensional network this is necessary.
2. Concentration by evaporating the water of milk may be increase the stabity of the foam, because the casein micelles may be larger in size after evaporation, and increased its concentration as well as increased its viscosity that is just as it should be increased the stability of the foam
3. Milk fat globules may also be penetat in the air-milk interface , while the milk in whipping, a lot of globules gathered at the surface of air bubbles and stick to the air-milk interface.The milk fat clusters with the air bubbles and fat globules form a network and make a rigid structure bubbles. If there are no sufficient milk fat to make rigid structural air bubbles, we can't get a stable foam, therefore the content of milk fat in the raw milk is necessary.
4. Season and lactation period also influence the foam stability, with the increse concentration of Ca ion, the quick of foaming lead to, and get a firm foam.

Factors affect the creaming

1. Generally, the flocculation of milk fat globules increase the creaming and rapid creaming occur. The flocculation by a special lactoglobulin is distinct improvement, and the special globulin content in raw milk is related to season, lactation period, feed and mineral [illegible]tion in milk.
2. The higher fat content and the larger fat globules in the milk, the rapid creaming is obtained, these are related to the breed of cow, in this reson Mongolian cow is suitable for WURUM manufature.
3. The change of ion concentration also influence the fluccolation.
4. Temperature effect is notable.

STUDIES ON PHYSIOCHEMICAL COMPOSITION OF GOATS' MILK

CHENGXIANG LUO MINGRUO GUO

MATERIALS and METHODS**

Milk, from 30 2-4 year-old Xinong Saanen goats (XSG) weighing 50-60kg, 15 4 year-old English Saanen goats (ESG) weighing 50-60 kg and 16 2-3 year-old Crossbred goats (CG = Xinong Saanen × native goat) weighing 40-50 kg was tested. Chemical components of the milk for the whole lactation (Colostrum, milk, and dry milk) were determined comprehensively. Some physiochemical characteristics were also studied. All results were statistically tested.

RESULTS*	COLOSTRUM			NORMAL MILK			DRY MILK		
	XSG	ESG	CG	XSG	ESG	CG	XSG	ESG	CG
SPECIFIC GRAVITY (20°C/4°C)	1.0377	1.0388	1.0354	1.0291	1.0276	1.0276	1.0351	1.0338	1.0347
pH	6.62	6.56	6.70	6.72	6.72	6.68	7.27	7.48	6.93
ACIDITY (°T)	20.70	20.95	16.94	11.46	10.03	11.26	9.8	6.56	11.99
FAT (%)	6.25	5.13	6.03	5.68	3.81	3.79	4.92	5.58	5.65
LACTOSE (%)	3.99	4.05	4.26	4.58	4.61	4.62	1.89	1.78	2.90
PROTEIN (%)	7.21	6.41	6.03	3.54	3.29	3.74	6.91	7.66	6.09
TS %	20.39	16.51	18.26	12.58	12.39	12.87	16.79	18.74	17.06
ASH %	0.99	0.82	0.86	0.86	0.79	0.88	1.29	1.29	1.21
Ca (mg/100g)	353.00	260.85	345.08	213.71	210.47	226.95	228.52	234.78	278.58
P (mg/100g)	157.05	116.36	135.74	96.18	85.34	99.42	125.94	126.36	135.95
Na (PPm)	611.08	670.72	637.1	473.36	426.75	447.99	>1300	>1300	>1300
K (PPm)	1992.98	2105.25	1749.80	2135.50	2141.95	2253.97	1194.35	898.57	1192.8
Mg (PPm)	168.60	146.13	149.50	148.68	139.66	147.11	235.07	290.87	222.6
Zn (PPm)	6.3999	6.698	5.1987	3.7864	3.5723	3.9976	4.535	5.272	4.574
Fe (PPm)	2.2849	3.1929	2.8944	1.4281	1.6519	1.6391	6.635	4.2066	2.881
Cu (PPm)	1.5296	2.3208	1.3556	0.5604	0.495	0.5751	1.6113	1.2247	0.512
Mn (PPm)	0.0983	0.0911	0.1081	0.0732	0.0646	0.0720	0.1815	0.1881	0.1669

* 1) Figures in this table are average value.
2) FA, AA in total protein (casein, whey protein), and FAA were omitted (AVAILABLE UPON REQUEST)

** INTRODUCTION OMITTED (AVAILABLE ON REQUEST)

PHYSIOCHEMICAL COMPONENTS	METHODS / INSTRUMENTATION
SPECIFIC GRAVITY, ACIDITY, PH, TS, ASH	AOAC METHODS
Ca / p	POTASSIUM PERMANGANATE OXIDATION-REDUCTION TITRATION / AMMONIUM PHOSPHO-MOLYBDATE CALORIMETRIC METHOD
LACTOSE	MICRO-CALORIMETRIC ANALYSIS
TOTAL PROTEIN	SIMPLE MICRO KJELDAHL TEST
AMINO ACIDS (AA)	AMERICAN 121 MB AMINO ACIDS ANALYZER
Mg, Zn, Na, K, Fe, Cu, Mn	ICPQ-1000 INDUCTIVELY COUPLE PLASMA ATOMIC EMISSION SPECTROSCOPY
FATTY ACIDS (FA)	GAS CHROMATOGRAPHY (GC-7A TYPE)

CONCLUSION:

I. THE DIFFERENCES BETWEEN MOST OF THE CHEMICAL COMPONENTS WERE SIGNIFICANT ESPECIALLY WHEN COMPARING COLOSTRUM AND NORMAL MILK, COLOSTRUM AND DRY MILK, AND NORMAL AND DRY MILK. IN THE SAME STAGE OF LACTATION, THE DIFFERENCES BETWEEN SOME OF THE CHARACTERISTICS OF TWO BREEDS WERE ALSO SIGNIFICANT.

II. DURING THE WHOLE LACTATION PERIOD (COLOSTRUM-NORMAL MILK-DRY MILK) THE PHYSIOCHEMICAL COMPOSITION OF THE MILK FROM ALL THREE BREEDS HAD THE SAME TREND OF CHANGES AS COW'S MILK.

III. IN DRY MILK, IT IS FOUND THAT THE pH VALUE AND THE CONTENTS OF PROTEIN AND ASH REACHED THE HIGHEST LEVEL OF THE WHOLE LACTATION, AND TENDED TO HAVE SIMILAR CONTENT AS CORRESPONDING COMPONENTS IN BLOOD, WHILE THE ACIDITIES AND THE LACTOSE CONTENTS REACHED THE LOWEST LEVEL. THESE STUDIES SUPPORT THE OPINION THAT IN SOME CASES MILK FAT AND PROTEIN CURVES MAY CROSS EACH OTHER DURING LATE LACTATION.

IV. THE RATIO OF Glu/His IN CASEIN (7.63, 7.90, AND 7.19) IS LOWER THAN THAT OF COW'S MILK (9.13). THE TOTAL FAA CONTENT IS MUCH HIGHER THAN THAT OF COW'S MILK.

V. THE RATIO OF Na/K IN GOAT'S MILK (ABOUT 0.20) IS SIGNIFICANTLY LOWER THAN THAT OF COW'S MILK (ABOUT 0.31). THIS MAY ACCOUNT FOR THE LOWER STABILITY OF GOAT'S MILK IN THE ALCOHOL TEST.

VI. THE ACIDITY, LACTOSE, Na, Fe, Zn, Cu IN GOAT'S MILK ARE LOWER THAN THOSE IN COW'S MILK AND TS, ASH, K, Mn AND Mg ARE HIGHER.

Northeast Agricultural College • Harbin, Heilongjiang • CHINA

CONTINUED MONITORING OF IODINE LEVELS IN FRESH PASTEURISED MILK FOR THE CITY OF ADELAIDE, SOUTH AUSTRALIA (POPULATION 925,050)

B.D. HANNAFORD
METROPOLITAN MILK BOARD
33 HUTT STREET
ADELAIDE
SOUTH AUSTRALIA 5000

1. As a follow on from the Brief Communications 21st International Dairy Congress, Moscow U.S.S.R. 1982, samples of pasteurised milk have been continued to be submitted for iodine analysis.

2. A summary of the results is tabulated.

IODINE LEVELS IN MILK

	No. of Samples	Av. µg/l Iodine	Range µg/l
1/7/80 - 30/6/81	28	210	110 - 400
1/7/81 - 30/6/82	32	250	120 - 460
1/7/82 - 30/6/83	27	270	160 - 520
1/7/83 - 30/6/84	22	200	<100 - 440
1/7/84 - 30/6/85	12	180	150 - 230

3. A standard of not more than 500 micrograms per litre is recommended by the Australian National Health and Medical Research Council.

4. This standard has been adopted by the South Australian Health Commission's Food Standards Code 1985.

5. The Metropolitan Milk Board through face to face communication with its milk producers (farmers) has continued to press home the proper use of sanitizers. This action has been supplemented with precise literature.

6. A warning label stating "This product is not to be used for dairy cows when iodine residues are a problem", is now compulsory on mineral supplements containing iodine. This action has been promulgated by the Stock Medicines Board of South Australia.

7. The Metropolitan Milk Board intends to continue to monitor iodine levels in pasteurised milk.

REFERENCE

1. Hannaford, B.D.: XXI INT. Dairy Congress Vol. 1, book 2, p. 576 (1982).

ANALYTICAL ERROR INVOLVED IN THE MILK SOLID CONTENT MEASURED BY HEAT DRYING METHOD AND A PROPOSAL FOR THE IMPROVEMENT

Takatoshi Itoh and Susumu Adachi
Faculty of Agriculture, Tohoku University, Tsutsumidori
Amamiyacho 1-1, 980 Sendai, Japan.

Milk solid content are usually measured by drying of milk in an air oven near 100°C, and such method is used as an official standard method for the determination of milk solid and used occasionally as a standard for calibration of the other analytical technique. The solid content obtained by heat drying method involves, however, some analytical error arising from heat induced degradation of milk components. In this report, the source and extent of analytical error involved in milk solid content determined by heat drying method were examined.

Table 1 shows the changes in milk solid content and remaining moisture content in the solid measured after repeated heating. The solid content of milk continuously decreased by repeated heating and constant weight was difficulty attained. The residual moisture in the milk solid reached minimum after 3 h heating, but about 1 % moisture was still remaining in it.

Volatile components such as carbonyls, hydrogensulfide, acidic and basic compounds were detected generating from milk solid during heating. This indicates that a part of milk constituents decompose during heating and it results in the continuous loss of weight. The differential thermal analysis revealed that lactose in milk solid is dried in the anhydrous form. By freeze drying, milk was well dehydrated into porous state without any sign of degradation. Moisture content remaining in the dried solid was less than 0.5 % (Table 2). This method is recommendable as a standard drying procedure.

Table 1. Changes in milk solid content and residual moisture content in dried solid after heating at 99 ± 1°C in air oven.

Heating time (h)	Milk solid content (%)	Moisture remaining in dried solid (%)
1	11.87±0.07 (n=10)	2.20±0.67 (n=10)
2	11.77±0.05 (n=10)	1.47±0.43 (n=10)
3	11.70±0.03 (n=10)	0.98±0.10 (n=10)
4	11.66±0.04 (n=10)	1.02±0.12 (n=10)
5	11.64±0.04 (n=10)	0.98±0.14 (n=10)
6	11.62±0.03 (n=10)	0.96±0.12 (n=10)
20	11.50±0.03 (n=10)	0.71±0.16 (n=10)

Table 2. Milk solid content and residual moisture content in dried solid after freeze and heat drying.

Method	Drying time (h)	Milk solid content (%)	Moisture remaining in dried solid (%)
Freeze drying	4	11.69±0.05 (n=5)	0.54±0.14 (n=10)
	5	11.68±0.01 (n=5)	0.53±0.11 (n=10)
	6	11.68±0.05 (n=5)	-
Heating method (99±1°C)	3	11.79±0.03 (n=5)	1.08±0.22 (n=10)

COMPOSITION AND PHYSICOCHEMICAL PROPERTIES OF MASTITIC MILK WITH THE CONSIDERATION OF ITS SUITABILITY FOR PROCESSING

W. Sajko, J. Kisza, P. Przybyłowski, B. Staniewski, J. Urbańska
Institute of Dairy Technology, Academy of Agriculture and Technology, Olsztyn, Poland.

Chemical composition, physicochemical properties and technological suitability of normal and mastitic milk were compared. Unfavourable changes in chemical composition of mastitic milk negatively affected the technological suitability of this milk. The internsification of these changes was dependent on the degree of mastitis development.

A COLORIMETRIC ENDOTOXIN ASSAY FOR THE ASSESSMENT OF THE BACTERIOLOGICAL QUALITY OF MILK

J. MOTTAR & M. NAUDTS

SCOPE

rapid estimation of gram-negative bacteria in milk and milkproducts

PRINCIPLE OF THE METHOD

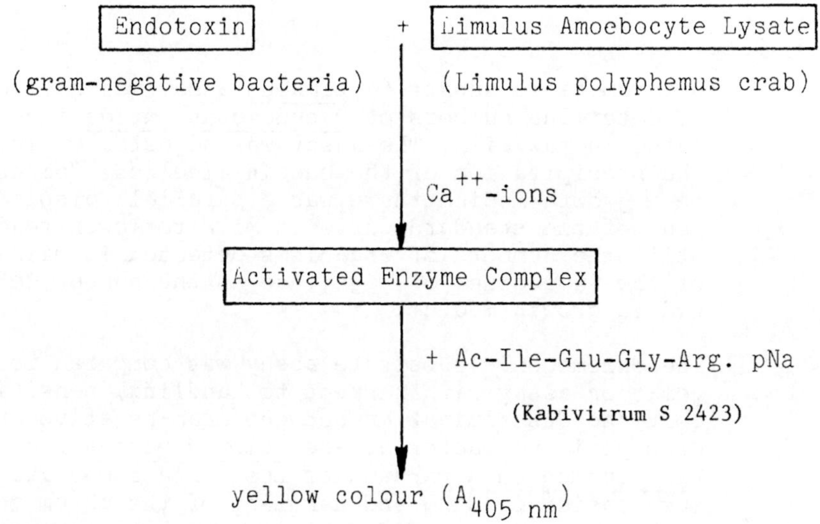

CONCLUSIONS

- the assay is rapid ($<$ 1 h), sensitive ($<$ 100 gram-negatives/ml) and accurate
- the endotoxin content is significantly correlated ($P < 0{,}01$) with the gram-negative bacterial count
- the test may be applied to the screening of raw materials, to raw milk grading and to in-line quality control monitoring

Governmental Dairy Research Station, 9230 Melle, Belgium

A CHROMOGENIC SUBSTRATE LIMULUS ASSAY APPLIED TO RAW MILK

Bärbel Hahn-Hägerdal and Anneli Svensson

Applied Microbiology, Chemical Center, Lund University, Box 124, S-221 00 Lund, Sweden.

A chromogenic substrate *Limulus* assay was applied to determine numbers of *Pseudomonas putida* inocculated in raw milk. The assay was adjusted to include the precipitation of the casein micelles. Compared to a growth medium there was a parallell displacement of the standard curve in milk to lower readings. Still the numbers of organisms detected in milk were of the same order of magnitude as the number detected in growth medium.

The chromogenic substrate assay was compared to a gelation assay with respect to handling, sensitivity, cost and discrimination between Gram-negative and Gram-positive bacteria. *Pseudomonas putida* and *Bacillus cereus* were used as test organisms. The much easier reading and handling of the chromogenic substrate assay should make it suitable for screening of large numbers of milk samples.

FOULING OF MEMBRANE DURING ULTRAFILTRATION OF LACTIC FERMENTED MILK

R.S. Patel*, H. Reuter
Federal Dairy Research Centre, Kiel D-2300, Postfach 6069, F.R.G.
*National Dairy Research Institute, Karnal 132001, India

With a decreased pH of milk the fouling process became faster, especially below pH 5.6. Also the final relatively steady flux was appreciably lower with lower pH. It seemed that aggregation of casein with decreasing pH enhanced fouling of the membrane. Presence of fat caused faster fouling, however, the extent of fouling was not significantly influenced. Decrease in flux was appreciably faster with skimmilk than with whey.

Deposit formation on a ultrafiltration membrane during the concentration of lactic fermented milk

R.S.Patel and H. Reuter
Institute für verfahrenstechnik,
der Bundesanstalt fur milchforschung
kiel, West-Germany

Ultrafiltration of lactic fermented milk using a hollow fibre membrane module was carried out in order to find out effects of pH and temperature of milk on the deposit formation or permeate flux, as well as to determine the composition of membrane deposits. Increasing concentration of solids resulted in continuously decreasing flux, the decrease being faster at low pH. pH value of 5.6 and lower caused reduction in flux to the extent that it was impossible to go beyond a volume reduction of 40 to 45% as against about 82% volume reduction with normal milk pH. The flux was greater at high temperature, however, the difference between 48°C and 17°C was much less at the end of concentration then in the beginning.

Decreased pH of milk resulted in increased protein content of the deposit and decreased ash, calcium and phosphorous content. Polyacrylgel electrophoresis of the protein materials in the deposit revealed that the fraction found in deposit were the same as found in normal casein.

EFFECTS ON THE PERMEATION OF SOLUTES FROM SKIMMILK AND ULTRAFILTRATION PERMEATE THROUGH REVERSE OSMOSIS MEMBRANES

Dr.-Ing.Ulrich Kulozik and Prof.Dr. H.G. Kessler

Institute for Dairy Science and Food Process Engineering, T.U.Munich,
D-8050 Freising-Weihenstephan

INTRODUCTION

THE PERMEATION OF SOLUTE J_S THROUGH REVERSE OSMOSIS MEMBRANES IS AN IMPORTANT CRITERIUM FOR EVALUATION OF THE PROCESS. BESIDES THE MEMBRANE MATERIAL SOLUTE PERMEATION IS HEAVILY INFLUENCED BY THE COMPOSITION OF THE SOLUTION CONCENTRATED. THE OBJECTIVE WAS TO STUDY THE PERMEATION OF SOLUTE (LACTOSE, SALT IONS) FROM ULTRAFILTRATION PERMEATE AND SKIMMILK AND TO FIND OUT CRITERIA FOR THE MEASUREMENT OF THE PERMEATE QUALITY.

MATERIALS AND METHODS

PILOT PLANT DESCRIPTION: FOR THE EXPERIMENTS A PILOT PLANT WITH TUBULAR MEMBRANES FROM PCI, WHITCHURCH, GB WAS USED. THE MEMBRANE TYPE USED HAS BEEN ZF 99 POLYAMIDE/POLYSULFONE.

MEASUREMENT OF LACTOSE: THE LACTOSE CONCENTRATION IN THE PERMEATE WAS MEASURED ENZYMATICALLY. REGARDING THE SOLUTION - DIFFUSION - MODEL THE SOLUTE PERMEATION WAS EXPRESSED BY THE EQUATIONS

$$J_S = B\,(c_2 - c_3) = J_{H_2O}\,c_3$$

RESULTS: INFLUENCE OF THE FLUID COMPOSITION ON THE SOLUTE PERMEATION

EFFECT OF IONS AND PROTEIN LAYERS AN LACTOSE PERMEATION

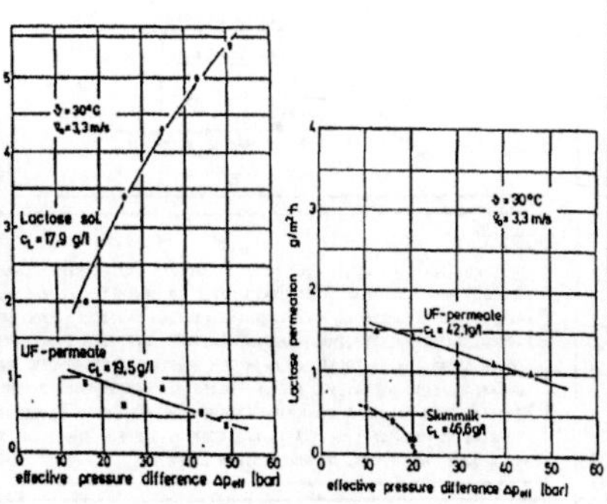

EFFECT OF DRY MATTER CONTENT ON PERMEATE QUALITY

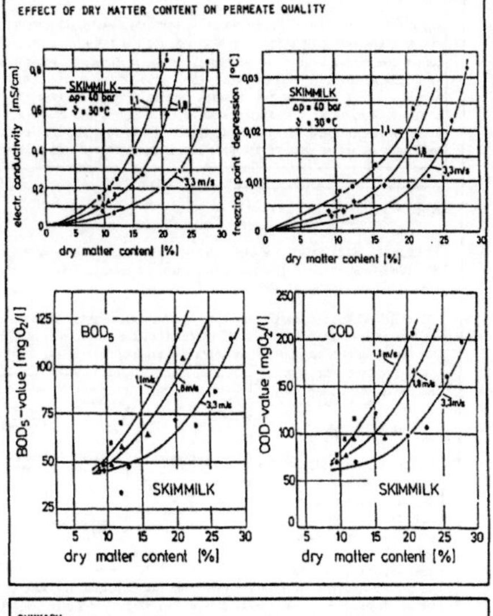

SUMMARY:

THE EFFECT OF FLUID COMPOSITION AN SOLUTE PERMEATION WITH REVERSE OSMOSIS OF SKIMMILK AND UF-PERMEATE WAS INVESTIGATED. IT COULD BE FOUND THAT THE LACTOSE PERMEATION IS REDUCED BY IONS AS A RESULT OF ELECTRICAL INTERACTIONS OF THE IONS WITH THE MEMBRANE SURFACE. PROTEIN LAYERS PARTIALLY ACT AS A SOLUTION - DIFFUSION - MEMBRANE AND REDUCE THE PERMEATION OF IONS AND LACTOSE. THE PERMEATE QUALITY DECREASES SHARPLY WITH INCREASING CONCENTRATION RATIO WHAT CAN EASILY BE MEASURED BY THE ELECTRICAL CONDUCTIVITY OF THE PERMEATE OR BY THE FREEZING POINT DEPRESSION.

INSTITUTE OF DAIRY SCIENCE AND FOOD PROCESS ENGINEERING
TECHNICAL UNIVERSITY OF MUNICH, D-8050 FREISING - WEIHENSTEPHAN

EFFECTS ON THE FORMATION AND RINSING BEHAVIOUR OF PROTEIN DEPOSITS FROM SKIMMILK ON REVERSE OSMOSIS MEMBRANES

Dr.-Ing.Ulrich Kulozik and Prof.Dr. H.G.Kessler
Institute for Dairy Science and Food Process Engineering, TU Munich, D-8o5o Freising-Weihenst.

INTRODUCTION

THE FORMATION OF PROTEIN DEPOSITS ON RO MEMBRANES HAS GREAT INFLUENCE ON THE CONCENTRATION PROCESS ITSELF AS WELL AS FOR THE MEMBRANE CLEANING. DURING CONCENTRATION MOST OF THE PROTEIN MATERIAL SHOULD BE PREVENTED FROM FORMING DEPOSITS BY HIGH AXIAL FLOW FORCES TO GET A HIGH PERMEATION RATE. IN THE BEGINNING OF THE CLEANING PROCESS MOST OF THE DEPOSIT SHOULD BE REMOVED BY A RINSING PERIOD TO MINIMIZE THE ENERGY AND AGENTS REQUIRED FOR CHEMICAL CLEANING. THE OBJECTIVE WAS TO STUDY THE INFLUENCES ON FORMATION AND REMOVAL OF DEPOSITS DURING CONCENTRATION AND RINSING.

MATERIALS AND METHODS

PILOT PLANT DESCRIPTION: FOR THE EXPERIMENTS A PILOT PLANT WITH TUBULAR MEMBRANES FROM PCI, WHITCHURCH, GB WAS USED. THE MEMBRANE TYPE USED HAS BEEN ZF 99 POLYAMIDE/POLYSULFONE.

ESTIMATING DEPOSIT REMOVAL: PREVIOUSLY TO THE EXPERIMENTS THE WATER FLUX OF THE CLEAN MEMBRANE WAS MEASURED UNDER DEFINED CONDITIONS (M_{H_2O} ①) AND COMPARED WITH THE WATER FLUX AFTER THE RINSING PROCESS (M_{H_2O} ②) UNDER THE SAME CONDITIONS. THE RATIO

$$M_{rel} = \frac{M_{H_2O} ②}{M_{H_2O} ①}$$

WAS TAKEN AS MEASURE OF PROTEIN REMOVAL. TWO SERIES OF EXPERIMENTS WERE RUN: THE FIRST WITH CONSTANT CONDITIONS DURING RINSING, THE SECOND WITH CONSTANT CONDITIONS DURING CONCENTRATION. IN BOTH THE EFFECT OF TRANSMEMBRANE PRESSURE DIFFERENCE, FLOW VELOCITY AND COMPOSITION OF THE PRODUCT HAVE BEEN INVESTIGATEI

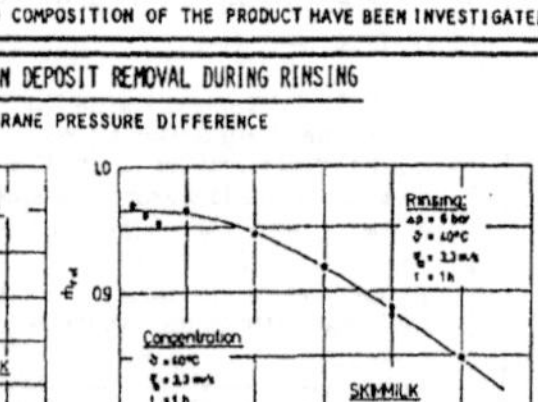

RESULTS: INFLUENCES ON DEPOSIT REMOVAL DURING RINSING

EFFECT OF THE TRANSMEMBRANE PRESSURE DIFFERENCE

EFFECT OF THE AXIAL FLOW VELOCITY

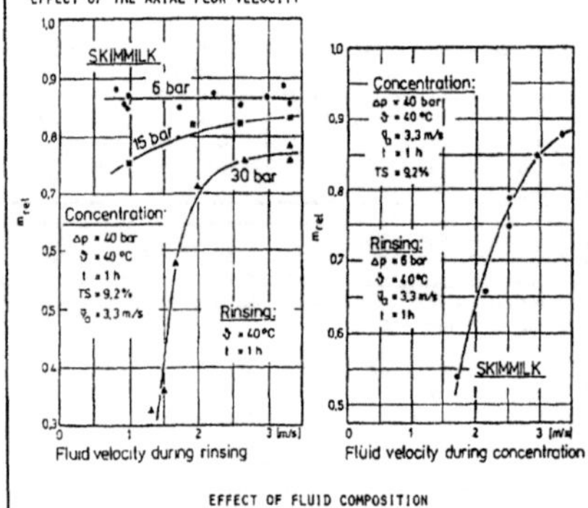

EFFECT OF FLUID COMPOSITION

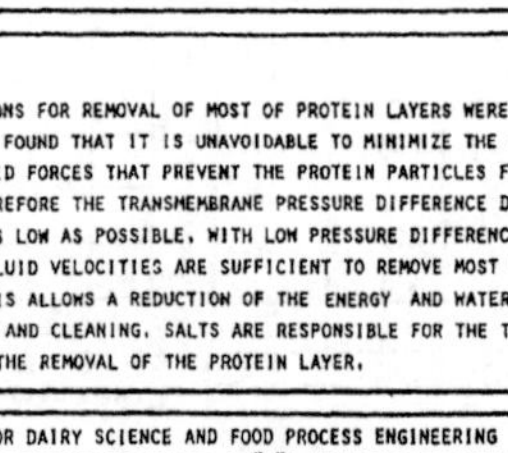

SUMMARY

THE CONDITIONS FOR REMOVAL OF MOST OF PROTEIN LAYERS WERE EXAMINED. IT COULD BE FOUND THAT IT IS UNAVOIDABLE TO MINIMIZE THE RADIAL COMPACTING FLUID FORCES THAT PREVENT THE PROTEIN PARTICLES FROM BEING RINSED. THEREFORE THE TRANSMEMBRANE PRESSURE DIFFERENCE DURING RINSING SHOULD BE AS LOW AS POSSIBLE. WITH LOW PRESSURE DIFFERENCES, RELATIVLY LOW AXIAL FLUID VELOCITIES ARE SUFFICIENT TO REMOVE MOST OF THE PROTEIN DEPOSIT. THIS ALLOWS A REDUCTION OF THE ENERGY AND WATER REQUIREMENTS FOR RINSING AND CLEANING. SALTS ARE RESPONSIBLE FOR THE TIGHT STRUCTURE AND REDUCE THE REMOVAL OF THE PROTEIN LAYER.

INSTITUTE FOR DAIRY SCIENCE AND FOOD PROCESS ENGINEERING
TECHNICAL UNIVERSITY OF MUNICH, D-8o5o FREISING-WEIHENSTEPHAN

Ultrafiltration of Soluble and Heat Precipitated Whey Protein Concentrates

D. Roehl and P. Jelen
Department of Food Science
University of Alberta
Edmonton, Alberta, T6G 2P5
Canada

Recent studies (1) have indicated the potential for flux improvements in ultrafiltration of isoelectrically precipitated soy proteins in comparison to soluble soy proteins. Significantly different transmembrane-pressure-drop vs permeate flux profiles, and protein content vs flux profiles of the soluble and precipitated systems were related to flux improvements at high protein concentrations. The ultrafiltration of previously UF-prepared 50 litre aliquots of heated (90°C, 15 minutes) 1.2% protein retentates (12% TS) and diafiltrates (5.5% TS) from cottage cheese whey in a DDS Lab-20 UF unit equipped with 20 GR61-PP polysulphone membranes did not result in improvement of UF performance compared to identical soluble systems. In conditions of 200-1100 kPa and 20°C, fully heat denatured and precipitated protein systems did not result in higher flux rates for a given applied transmembrane pressure. Precipitated whey proteins, when concentrated batchwise, exhibited a more rapid drop in permeate flux and at a lower TS content than soluble whey proteins. At 17.5% TS, near the maximum concentration of 18% TS obtained in the diafiltered precipitated system, the measured viscosity was 230 centipoise. This compared to 45 centipoise for a 17.5% TS diafiltered soluble concentrate. The lower viscosities of the soluble proteins allowed final concentrations of 30% and 40% TS for soluble diafiltered and high lactose systems respectively. It was observed that diafiltered WPC exhibited shear thinning at 25% TS and 11% TS for soluble and precipitated protein systems respectively.

References

1. Hoare, M. and Devereux, N. 1985. "Membrane Recovery of Soluble and Precipitated Soya Proteins". 4th Int. Congress on Engineering and Food, Edmonton, July 7-10, 1985. Digest of Papers, p. 141.

DROPLET SIZE IN SPRAYS OF ATOMIZED CONCENTRATED MILK

P.J.J.M. van Mil, H.J. Klok, A.J. Damman
Netherlands Institute for Dairy Research, P.O. Box 20,
6710 BA EDE, The Netherlands

Droplet size distribution of concentrated milk which was atomized by nozzle or wheel was measured. Experiments were carried out to resolve the effect of gaspressure, rotational speed and properties of the concentrated milk. The results hardly correlate with the equations for the median of droplet size distribution as proposed in literature. They confirm the importance of the viscosity of the concentrate and question an effect of sweet-cream buttermilk.

WEDNESDAY - POSTER 27

RECOMMENDATIONS FOR MINIMIZING PRODUCT LOSSES IN EVAPORATORS*

S. Bouwman, A.E. Steenbergen, R. Waalewijn
Netherlands Institute for Dairy Research (NIZO), P.O. Box 20, 6710 BA EDE, The Netherlands

The operating conditions of several evaporators for whole and skim milk used in the Dutch dairy industry were investigated. Recommendations are given to minimize the total evaporation costs, including investment- and energy costs, and also the costs of product losses and waste water disposal. By taking the adequate measures the product losses of modern evaporators can be limited to 0.4 per cent of the capacity for whole milk and for skim milk to 0.2 per cent or even lower.

*Complete poster available from authors.

THE VISCOSITY OF WHOLE MILK CONCENTRATES

M. Hallström and P. Dejmek
Department of Food Technology, University of Lund, P.O. Box 124, S-221 00 Lund, Sweden

Whole milk was concentrated by ultrafiltration to the concentration levels of 1, 3, 4, 4.5 and 5 times the original concentration. The viscosity of the different concentrates was measured as a function of the temperature (temperatures from 20°C to 50°C) and as a function of the pH-value (pH-values from 5.2 to 6.7) at a shear rate of 466 s^{-1}. The viscosity measured was correlated to the voluminosity of the casein micelles given by Snoeren et al.

WEDNESDAY - POSTER 29

IMPROVING THE DISPERSIBILITY OF MILK POWDER

Liquan Huang
Heilongjiang Dairy Industry Research Institute,
113 Xuefu Road, Harbin, P.R. China.

Dispersibility test of pnematic-separated fractions from a batch of milk powder and technology experiment in pilot plant all indicated that properly increasing particle size is a realistic way for making warm water "instant" milk powder. For making cold water "instant" milk powder, lecithination on fluid bed or within spray dryer seems to be similarly effective.

EFFECT OF ALKALI-TREATED RICE STRAW-MANURE SILAGE ON MILK YIELD, MILK COMPOSITION OF HOLSTEIN COWS

NAM-HYUNG LEE. KOREA ADVANCED INSTITUTE OF SCIENCE AND TECHNOLOGY. P.O.Box 131, DONG DAE MUN, SEOUL, KOREA

SUMMARY This experiments were carried out to evaluate the effects of feeding straw-manure silage compared with corn silage, and were consisted of trial I by Halstein cow and trial II by sheep with change-over feeding. And trial II was used to evaluate the digestibility and ruminal characteristics. The ratio of formula feed to each silage during the experimental period was maintained at 40:60 (w/w) as DM basis.

Data from trial I on corn and straw-manure silage group, respectively, were; daily milk yield 17.23, 17.05kg; milk fat 4.06, 4.19%; milk protein 3.89, 3.92%; milk lactose 5.27 5.17%; SNF 10.36, 10.38%. There values were not significantly different between treatments.

The digestibility of ration components on trial II showed that the digestibility of proximate composition of corn silage tended to be slightly increased. But the fibrous material digestibility of straw-manure silage showed more increased results ($P < 0.05$). Ruminal parameters on pH and VFA concentration were similar with two silage group. However, straw-manure silage resulted in stepwise increase in propionate and valerate throughout the sampling time.

INTRODUCTION Corn is the most popular cereal which is conserved as silage and large areas are grown for this purpose in many parts of the world. Corn silage has always been regarded as an important group of roughage for feeding dairy cows in the Korea during winter.

A major disadvantage of corn silage is, however, its low crude protein content, usually being less than 100g kg^{-1} DM. And this silage is not always sufficient for feeding dairy cows because of the limited grass-land and the amount produced.

It is, therefore, logical to try to utilize as much as possible of feed resources available in our country, such as agricultural by-products.

In recent years, the application of chemical-treated roughage has an increasing effect of feed intake and digestibility (Klopfens tein, 1978). An animal waste contains consideral quantities of protein (12 to 25%) and carbohydrate (40 to 70%) that could provide useful nutrients when fed to ruminants (Bucholtz et al., 1971).

At KAIST, we developed a process to produce a silage using alkali-treated rice straw, animal waste and bran materials. The present product, so called "straw-manure silage", contains about 12 to 15% crude protein, 55 to 60% TDN value, low pH and high lactic acid content.

This study was carried out to evaluate the effects of feeding straw-manure silage as compared with corn silage.

MATERIALS AND METHODS The rice straw, chopped into 3-5cm lengh, was treated with 4% NaOH by dry weight of straw and consequently, fixed to 50% moisture content. The mixing ratio of straw, manure and defatted rice bran was 50:20:30 as DM basis.

Ten-Holstein cows in trial I were used in each of 2 period in continuous feeding trials with change-over feeding. Cows in each period were assigned to the standardization period at 15 days and the experimental period at 70 days. The ratio of formula feed to roughage during the experimental period was maintained at 40:60 (w/w) as DM basis, and the small amount of hay was offered at evening after the formula feed and silage had eaten up. Milk yield were recorded twice daily for each cow. Daily composite milk sample were taken five times at each period for fat, protein, lactose, and SNF determination by MILKO SCAN 104 (FOSS Electric).

Four corriedale ram in trial II, average body weight 30kg, were allocated to two treatments with change-over feeding. The feeding ratio of experimental ration was identical with that of trial I. Sheep were in a total collection digestion trial to determine the digestibility, and rumen liquids were sampled for determination of rumen parameters.

The total VFA were made by the steam distillation methods (Fenner and Elliot, 1963) and individual VFAs analyses (Erwin et al., 1961) were conducted by GLC. And the packing material was GP 10% SP-1200/1% H_3PO_4 on 80/100 chromosorb W AW.

RESULTS AND DISCUSSION Average dietary concentrations of crude protein as calculated from consumption data by Holstein cows were 14.50 and 15.40% for corn silage and straw-manure silage group.

Results of the trial I with Holstein cows are in Table 2. Intake of dry matter from straw-manure silage was significantly higher ($P<0.01$) and total DM consumption of straw-manure silage group was higher than of corn silage group ($P<0.01$). Harmond et al (1975) reported that a reason for the increase of DM intake for the litter-containing silages may have been the bulk density of the materials. In addition to manure, this silage was supplemented with 30% of defatted rice bran as DM basis. Wilkins et al (1971) found close positive correlations between intake and fermented quality as estimated from lactic acid.

No difference between two forage group were statistically significant for milk yield and milk constituents (Table 2). The data of digestion trial by sheep showed that the general digestibility of corn silage group tended to be slightly improved (Table 3). The utilization of fibrous materials for straw-manure silage group, however, showed more increased result ($P<0.05$). This results were associated with the effect of alkali-treatment on low quality roughage. And the digestible nutrient contents of straw-manure silage group were not inferior to that of corn silage group.

Table 1. Chemical composition and nutrient content of experimental ration (DM %)

Item \ Ration	Corn silage	Straw-manure silage	Hay	Formula feed
Dry matter	26.86	32.26	86.84	86.88
Crude protein	9.93	12.21	13.60	20.30
Crude fiber	26.89	24.33	34.39	7.24
NFE	54.26	34.38	42.15	61.24
pH	3.87	5.24		
Total VFA	5.68	14.37		
C_2	2.58	4.06		
C_3	0.14	1.82		
C_4	0.06	3.45		
C_i-5	0.21	0.15		
C_5	-	0.74		
Lactate	2.04	4.12		
Crude protein	14.49	15.40		
Crude fiber	19.93	19.12		
TDN	69.42	67.47		

Table 2. Effect of straw-manure silage on milking performance.

Item \ Treat.	Corn silage	Straw-manure silage	SEM
Feed intake (DM kg/cow/day)			
Formula feed	6.82	6.82	0.12
Silage	8.31	9.32	0.13**
Hay	1.73	2.13	0.11
Total	16.86	18.26	0.14**
Milk yield (kg/cow/day)	17.23	17.05	0.41
Milk composition (%)			
Fat	4.06	4.19	0.07
Protein	3.89	3.92	0.08
Lactose	5.27	5.17	0.05
SNF	10.36	10.38	0.22

The results of rumen parameters by sheep are listed in Table 4. The rumen pH throughout the post-feeding time was slightly lower with cornsilage group. The total VFA concentration between two group was similar. And the molar proportions of acetate for corn silage was higher and propionate, valerate ($P < 0.05$) were lower compared to straw-manure silage group throughout the sampling time.

Roffler et al (1978) reported that dietary concentration of protein (between 16.2% and 12.2% cp) had a positive influence on milk production. However, Chandler et al (1976) observed no difference in total milk yield between groups of cows fed ration with 12.5% and 15% cp.

Milk fat depressing diets produce a fall in rumen pH and change in the proportions of rumen volatile fatty acids, generally characterized by decreased proportions of acetate and sometimes of butyrate and by increased proportions of propionate and valerate (Storry, 1970).

Many information abailable suggests that energy intake, forage: concentrates value, type of forage and concentrates all can make a contribution to milk yield and composition. There is no close relationship between milk protein and ruminal propionate production (Youself et al., 1970; Rogers et al., 1982). An increase in the uptake of acetate and butyrate are recognized as precursors of the fatty acids, whereas an increased uptake of propionate or glucose may depress the concentration of blood plasma of acetate and β-hydroxybutyrate and depress slightly fatty acid synthesis de novo (Rood, 1979).

Table 4. Effect of straw-manure silage on rumen characteristics.

Time Sampling Time, (hr)	pH		Total VFA (mM/l)		Individual VFA (molar %)					
					C_2		C_3		C_4	
	C.	S.	C.	S.	C.	S.	C.	S.	C.	S.
0	6.96	6.84*	51.45	49.63	66.34	66.11	19.62	22.16	11.21	8.76**
1	6.37	6.59**	95.49	89.26	62.82	59.04	21.78	23.05	12.79	14.42
3	6.01	6.35**	97.14	97.29	60.08	57.12	23.90	25.35	13.45	13.84
6	6.07	6.31	92.44	87.02	62.63	59.35	24.29	28.88	11.14	9.33
9	6.38	6.61	76.01	71.69	63.94	61.99	23.26	27.46	11.04	7.93*

*C : Corn silage, S : Straw-manure silage

Thus, if all other nutrients are supplied in adequate amounts, the magnitude of response in milk yield and composition will depend on alterations in the amount and/or type of protein, fiber and carbohydrate in the basal rations and the genetic ability of the cow (Thomas, 1983; Kuhn et al., 1980).

Table 5. Daily feed cost per milk yield and profit evaluation by Holstein cow.

Item Treate	Corn-silage group	Straw-manure silage-group
Feed cost, (won)		
Formula feed	1,500	1,500
Silage	1,391.9	1,011.2
Hay	411.7	440.7
Total	3,303.6	2,951.9
Milk yield (kg/day)	17.23	17.05
Milk fat content(%)	4.06	4.19
Feed cost/kg milk yield	191.74	173.13
Gross profit/per head	3,378.2	3,926.1

As shown in Table 5, the feed cost per head required for kg milk production was estimated at ₩191.74 and ₩173.13, respectively. And on concerning the milk yield, fat content and feed cost, the more profit of ₩547.9 per head was obtained in the straw-manure silage group.

It was postulated that this silage was a suitable roughage for dairy cows at the 100% level of substitution for corn silage, and its use in practics has a profitable effect on cost.

REFERENCES (1) Bucholtz, H.F., H.E. Henderson, T.W. Thomas and H.C. Zindel. 1971. Dried animal waste as a protein supplement for ruminants. Proce. Internatl. Symp. on livestock wastes. ASAE. Publ. Proc. 271:308-310. (2) Chandler, P.T., C.A. Brown, R.P. Johnson, Jr., G.K. Macleod, R.D. McCarthy, B.R. Moss, A.H. Rakes, and L.D. Satter, 1976. Protein and methionine hydroxy analog for lactating cows. J. Dairy Sci. 59: 1897-1909. (3) Erwin, E.S., J. Marco and M. Emery. 1961. Volatile fatty acid analyses of blood and rumen fluid by gas chromatography. J. Dairy Sci. 44:1768-1771. (4) Fenner, H. and J. M. Elliot. 1963. Quantitative method for determining the steam volatile fatty acids in rumen fluid by gas chromatography. J. Anim. Sci. 22:624-627. (5) Harmon, B.W., J.P. Fontenot and K.E. Webb, Jr. 1975. Ensiled broiler litter and corn forage. II. Digestibility, nitrogen utilization and palatability by sheep. J. Anim. Sci. 40:156-160. (6) Klopfenstein, T. 1978. Chemical treatment of crop residues, J. Anim. Sci. 46:841-848. (7) Kuh, N.J., D.J., D.T. Carrick, and C.J. Wilde, 1980. Lactose synthesis : The possibilities of regluation. J. Dairy Sci. 63:328-336. (8) Roffler, R.E., L.D. Satter, A.R. Hardie, and W.J. Tyler. 1978. Influence of dietary protein concentration on milk production by dairy cattle during early lactation, J. Dairy Sci. 61:1422-1428. (9) Rogers, J.A., C.L. Davis, and J.H. Clark. 1982. Alteration of rumen fermentation, milk fat synthesis, and nutrient utilization with mineral salts in dairy cows. J. Dairy Sci. 65:577-586. (10) Rook, J.A.F. 1979. The role of carbohydrate metabolism in the regulation of milk production. Proc. Nutr. Soc. 38:309-314. (11) Storry, J.E. 1970. Reviews of the progress of dairy science. Section A. Physiology. Ruminant metabolism in relation to the synthesis and secretion of milk fat. J. Dary Res. 37:139-164. (12) Thomas, P.C. 1983. Milk protein, Proc. Nutr. Soc. 42:407-418. (13) Wilkins, R.J., K.J. Hutchinson, R.F. Wilson and C.E. Harris, 1971. The voluntary intake of silage by sheep. I. Interrelationships between silage composition and intake. J. Agric. Sci., Camb. 77:531-537. (14) Yousef, I.M., J.T. Huber, and R.S. Emery. 1970. Milk protein synthesis as affected by highgrain, low-fiber rations. J. Dairy Sci. 53:734-739.

LES POSSIBILITÉS D'INFLUENCER LA TENEUR DU LAIT EN PROTÉINES PAR L'ALIMENTATION

KRATOCHVÍL L., KASÝK M., ZADRAŽIL K.
(L'Universitaire de Lágriculture Prague, ČSSR)

Nous avons réussi à démontrer que la teneur du lait en protéines augmente de pair avec l'accroissement de la quantité d'amidon ingérée. Consécutivement à la distribution aux vaches laitières des collets et des pulpes de betterave ensilés ainsi que d'un mélange de collets et de pulpes ensilés la teneur du lait en protéines accusait les valeurs supérieures par rapport au mais ensilé distribué aux vaches laitières. La teneur la plus élevée (en p.c.) a pu etre démontrée lors de la distribution du fourrage vert avec les pulpes ensilés. Le mais ensilé accuse une teneur en protéines (en p.c.) inférieure meme avec la meme quantité de matière sèche comme dans le cas du mélange de pulpes et de collets ensilés ou bien des pulpes ensilées. La carence en matières azotées et en matières azotées digestibles est un facteur complémentaire.
Les résultats montrent que la teneur du lait en protéines la plus basse (en p.c.) est obtenue lors de la distribution du mais ensilé.
L'étude du niveau de matiére séche des rations alimentaires a révélé que la teneur du lait en protéines (en p.c.) allait de pair avec l'ingestion de matière sèche (de 13,9 à 13,7 kg) par les vaches laitières. Une augmentation prononcée de la teneur du lait en protéines est obtenue lors de la distribution des pulpes ensilées aussi bien pendant l'hiver que pendant la période d'été. C'est pourquoi il est préférable de distribuer les pulpes ensilées sous formes de composantes des rations alimentaires pendant toute l'année. La valeur d'amidon des rations alimentaires accuse une dépendance prononcée de la teneur du lait en protéines. La teneur du lait en protéines s'accroit de pair avec l'accroissement de la quantité absolute de valeur d'amidon ingérée par les vaches laitières (essais No.1 - 6,69; essai No. 2- 6,97). Les meilleurs résultats sont obtenus lors de la distribution du fourrage vert en combinaison avec les pulpes ensilées parce que dans ce cas les laitières ingèrent une quantité élevée d'unités d'amidon. Les essais n'accusent pas d'influence significative des matières azotées et des matières azotées digestibles sur la teneur du lait en protéines (en p.c.) lors d'une couverture suffisante des exigences des laitières (en p.c.).

L'INFLUENCE DE LA RASSE ET DE LA LACTATION SUR LA TENEUR DU LAIT EN ACIDES AMINÉS

KRATOCHVÍL L., MÁTLOVÁ J., ZADRAŽIL K.
(L'Universitaire de L'Agriculture Prague, ČSSR)

Nous avons choisi 40 primipares - 40 vaches croisées de la génération F_1 et F_{10} issues du croisement de la race tâchetée tchèque et de la race Holstein-Frisian. 20 laitières accusaient 25 % (CRC 25) et 20 laitières 50 % du sang de la race Holstein-Frisian. Les deux groupes accusaient une teneur élevée en acide glutamique suivie par la teneur en proline, en valine, en acide aspartique, en leucine et en lysine. La cystine n'a été démontrée qu'en traces. Des autres 16 acides aminés étudiés il y avait le moins de methionine, d'histidine et de glycine. Les deux groupes de laitières accusaient au cours du premier mois de lactation une teneur élevée an acides aminés par rapport à la moyenne pour toute lactation. L'abaissement le plus prononcé peut etre démontré dans le groupe CR 50 au cours du troisième mois de lactation, dans le groupe CRC 25 au cours du deuxième mois de lactation. Ensuite, la quantité totale d'acides aminés augmente au cours de chaque mois à l'exception du huitième mois où les deux groups de laitières accusent une chute d'acides aminés ainsi que de protéines. Dans le groupe CR 50 le niveau d'acides aminés atteint son maximum au cours du neuvième mois pour s'abaisser jusqu'à la fin de lactation, ce qui a trait au tarrissement des laitières. Dans le groupe CRC 25 la teneur en acides aminés à partir du neuvième mois reste pratiquement inchangée.
D'après nos résultats le groupe CR 50 de vaches croisées de la génération F_1 issues du croisement de la race tâchetée tchèque et de la race Holstein - Frisian accusait des valeurs plus élevées de teneur en acides aminés que le groupe CRC 50 qui a pris naissance grâce au croisement de retour avec un taureau tchèque tâcheté.
Nous avons réussi à démontrer que les deux groupes accusaient la varibilité la plus élevée pour la teneur en methionine et arginine. Les coéfficients de variabilité les plus bas ont été trouvés pour toute lactation dans les deux groupes de laitières pour la valine.
En conclusion, on peut dire qu'une variabilité supérieure a été prouvée pour toute lactation pour le groupe CRC 25.

UTILIZATION OF UREA/MOLASSES LIQUID FEED (LF) AS A MAJOR SOURCE OF NITROGEN AND ENERGY FOR LACTATING BUFFALO

K.M. Abdel-Rahman, B.M. Ahmed and A.N. Zedan*
Department of Animal Production, and *Dairy Science, Faculty of Agriculture, Monoufia University, Shebin El-Kom, Egypt.

Twelve lactating buffaloes were used in three feeding treatments in a switch-back design to study the effect of liquid feed containing urea, molasses, minerals and vitamins on milk production, composition and manufacture. Replacing concentrates by LF up to 40% was successfully used without adverse effects on milk yield, milk composition and manufacture.

WEDNESDAY - POSTER 34.

EFFECT OF ADDED SALTS AND HEAT ON THE PROPERTIES OF SUPERNATANTS OBTAINED BY MILK ULTRACENTRIFUGATION

Fatma A. Salama, A.M. Youssef, N.M. Abdel-Salam and A.E. Salam
Department of Agricultural Industry, Faculty of Agriculture, Alexandria University, Egypt.

The addition of calcium ions resulted in an increasing rate of nitrogen depletion, decreasing of the size and weight of casein micelles. However, addition of phosphate ions led to a marked decrease in the protein sedimentation rate. Heat combined with phosphate addition caused a further decrease in the rate of sedimentation, in size and weight of casein micelles.

Effect of season production and storage on the microbial coctent of ice - cream

R. M. Saleem and A. K. Abd El - Kadar
Food Technology Department , College of Agric. & Forestry, University of Mosul

Higher microbial content of ice - cream marketed in Mosul city were found during July and August . Lowest were found in May and September . This could be attributed to the high temperture in summer that enhance the rapid proliferation of bacteria . Also due to poor sanitation during processing.

Samples stored at -20 C for 10 weeks affected the total number of different types bacteria. The storage decreased the number of all types of the bacterial content . The rate of decreasing in bacterial number during the period of their storage . This may be related to the effect of refrigeration on the bacterial cell .

THE STUDY OF THE INFLUENCE OF THERMIC AND MECHANICAL PROCESSES ON THE MICROSTRUCTURE OF PROTEIN PHASE OF DAIRY PRODUCTS

R. Popko, H. Popko
Technical University, Lublin 1, skr. 189, Poland

In the article there have been presented the results of the study of the influence that homogenization process, separation and pasteurization may have on the microstructure of protein phase of milk and cream. The enclosed photographs of protein phase microstructures have been obtained through an electron microscope at the enlargement of about 10,000 times.

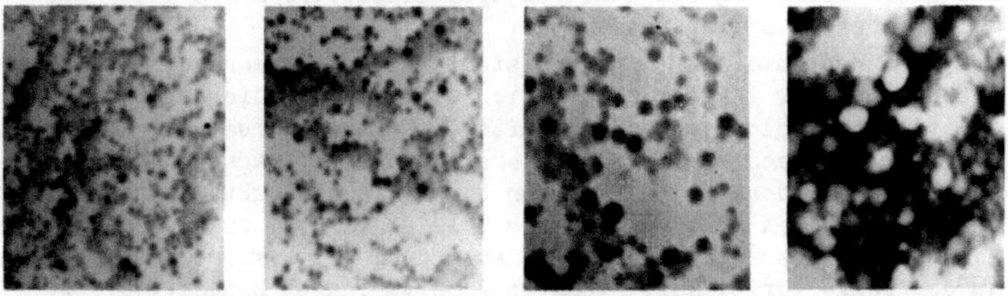

a b c d

Fig. 1. The microstructure of cream protein phase:
a - non-homogenized, b - homogenized at the pressure of 10,0 MPa, c - 15 MPa, d - 20 MPa.

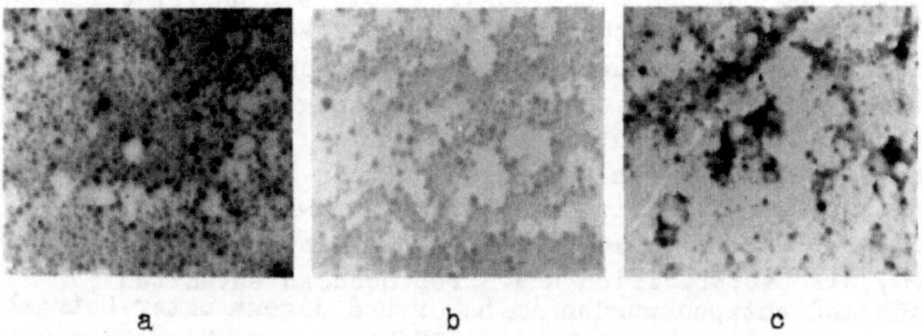

a b c

Fig. 2. The microstructure of milk protein phase
a - raw milk, b - centrifuged milk, c - pasteurized milk at the temperature of 82°C.

Die Problematik der Bakteriophagen und die Nutzung eines Laktomediums in der ČSSR

Ing. Jana Havlová, Ing. Ludmila Peterková und Koll.,
Forschungsanstalt der Molkereiindustrie, Prag

Obwohl in der ČSSR die Kontamination von Molkereibetrieben mit Bakteriophagen nicht die Hauptursache für die Inhibition von Milchkulturen ist, so tritt sie doch auf. Die Gruppe der Phagen, welche aus Betrieben stammen, wo es zu durch Bakteriophagen bedingten technologischen Schwierigkeiten kam, umfasst Phagen von Str. lactis, Str. thermophilus und Lbs. acidophilus.

Zur Unterdrückung der Phagenkontamination wird die empfindliche Kultur durch eine andere, deren Widerstandsfähigkeit gegenüber dem isolierten Phagen vorher geprüft wurde, ausgetauscht. Die Ersatzkultur wird aus einem Sortiment von Kulturen ausgewählt, welches ein auf die Herstellung von Kulturen spezialisierter Betrieb an die Molkereibetriebe liefert.

Die Kulturen, die als phagenresistent ausgewählt wurden, werden nach Beseitigung der Phagenkontamination wieder durch die ursprünglichen, meistens technologisch vorteilhafteren Kulturen ersetzt.

Eine der Massnahmen, die zur Unterdrückung der Phagenkontamination angewendet werden, besteht in der Nutzung eines Laktomediums, welches Salze enthält, die freies Calzium bilden, zur Züchtung von Milchsäurebakterienkulturen. Die Anwendung des Laktomediums ist auch in dem Falle vorteilhaft, wenn es sich um Phagen handelt, die zu ihrer Replikation keine Kalziumionen brauchen. Eine schnellere Säuerung,besonders in den ersten Phasen schränkt die Wirksamkeit der Bakteriophagen ein.

Laktomedium ist ein semiinstantes Nährmedium auf der Basis von Milch, Es enthält Bestandteile zur Stimulierung des Wachstums der Milchsäurebakterien. Weil es nicht möglich ist, ein universelles Medium für alle Arten der Milchsäurebakterien zur Verfügung zu haben und da die einzelnen Gattungen und Arten von Mikroorganismen sehr unterschiedliche Ansprüche an die Zusammensetzung des Nährmediums stellen, wurden zwei Nährmedien entwickelt.

Laktomedium S-für Kulturen von Streptococcus-Milchsäurebakterien gegebenenfalls in Kombination mit Leuconostoc-Bakterien.

Laktomedium L-für Lactobacillus-Kulturen einschliesslich von Mischkulturen, die Lactobacillen und Streptococcen enthalten.

Beide Nährmedientypen wurden im Labor und direkt unter Betriebsbedingungen in Molkereibetrieben zur Züchtung von Säureweckerkulturen geprüft und das auch in Betrieben, wo es zu technologischen Schwierigkeiten in folge von Phagenbefall kam.

THE INFLUENCE OF LACTATION PERIOD ON THE FATTY ACID COMPOSITION OF MILK FAT

DOC. RNDr.K.ZADRAŽIL ,CSc., PROF. ING.L.KRATOCHVÍL,DrSc.,
DOC. ING.J.PYTLOUN, CSc. (Agricultural University Prague,
ČSSR)

Chromatografic records were processed for the analyses of 750 milk fat samples (fatty acid methyl esters) obtained from different breeds of dairy cattle (the Bohemian Pied cattle prevailed).
The CHROM 4/1 chromatograph (produced by Laboratorní přístroje Praha, Czechoslovakia) was used for the analysis of milk fat; the length of the glass chromatographic column was 2 m and diameter 3.3 mm and the filling was 13 % DEGJ on Chromaton N-AW-DMCS, grain size 0.125/0.160 mm.

The effect of the stage of lactation (from the first hour post partum to the 50th day p.p.) was studied as exerted on the composition of the different fatty acids of milk fat. Special attention was paid to the intermediate period between colostrum and "ripe milk" production (the 6th to 9th day after calving); besides the major fatty acids of milk fat, emphasis was laid on the investigation of the minority fatty acids.

In all, 23 fatty acids of milk fat were determined, including 11 acids called minority fatty acids. The differences in the propotions of each fatty acid in colostrum and "ripe milk" were significant (P=0,01) or reached the margin of statistical significance. The differences in the content of fatty acids in the milk fat of the transient stage from colostrum to milk were mostly insignificant.

In this way it was demonstrated that the limit for colostrum can be decreased from the present 8 days after calving to 6 days after calving because milk and milk fat has almost the some composition in the "intermediate milk" and in "ripe milk".

POLYACRYLAMIDE GEL ELECTROPHORESYS OF WHEY PROTEINS IN CHEESES MADE FROM MILK OF DIFFERENT SPECIES

L. Amigo, G. Santamaria, D. Gonzalez de Llano, M. Ramos
Instituto de Fermentaciones Industriales, CSIC,
Juan de la Cierva 3, 28006 Madrid, Spain

Electrophoresis of whey proteins on polyacrylamide gels (1) was used to analyze cheese samples prepared with various proportions of ewes' and cows' milk and at various stages of cheese maduration. The three dimensional densitometric curves shown in Fig. 1 are for one month ripened cheese which were made from milk with ewes' milk with 0,5,10,20 and 100% cows milk.

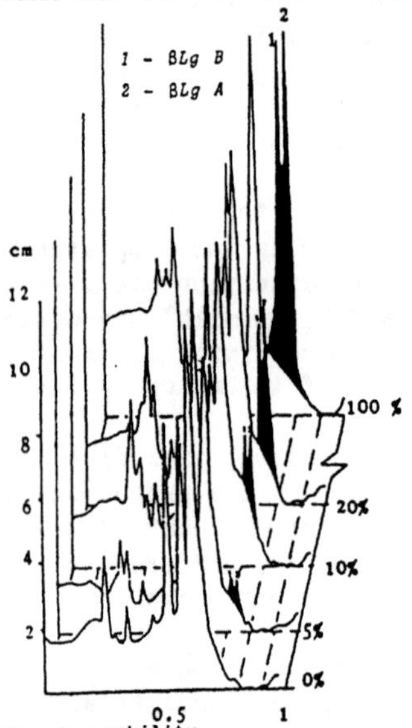

Fig. 1.- mobility

The curves indicated that β-lactoglobulin A and B have higher mobility for cows milk than for ewes milk. The β-lactoglobulin remains unaltered during the cheese ripening process (2).

The correlation coefficients and standard errors against the ripening time for different variables (β-Lg A/BSA, βLg A/α La, βLg A/βLg ewes, βLg B/BSA, βLg B/α- La, βLg B/ βLg ewes) were calculated. The low coefficients point to a low variability of the ratios over time. These remain essentially constant during cheese ripening. The high standard errors associated with each coefficient render these coefficients statistically insignificant.

Regression equation of the variable βLg B/BSA against the percentage of cows' milk used to make Manchego and Roquefort cheeses were obtained. A high correlation $> .99$ with low error $p < 0.001$ is obtained. It is evident that different regression equations are needed for different types of cheeses (Fig. 2).

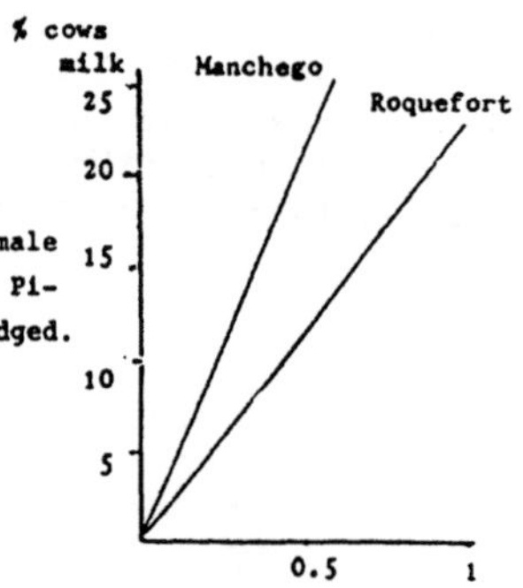

Fig. 2 -β-Lg B/BSA

ACKNOWLEDGEMENTS

The preparation of the samples of Roquefort cheeses by Mme Rivamale of the Roquefort Laboratories and the samples of Manchego by A. Pita of the Escuela de Industrias Lacteas are gratefully achnowledged.

REFERENCES

1. Hillier, R.M. 1976. J. Dairy Res. 43, 259-265.
2. Polo, C., Ramos, M., Sánchez, R. 1985. Food Chemistry, 16, 465-77

Biochemical studies on whey lipid substances.

R.S.Farag*;Y. Ghali* M.M.Hewedi**and O.M. Abdel-Fatah*.*Biochemistry Dept.,Faculty of Agriculture, Cairo University, Giza, Egypt ** Food Science Dept., Faculty of Agriculture at Fayoum, Cairo University, Fayoum, Egypt.

Introduction: Whey is a by-product of cheese and casein industry. In Egypt, the amount of the resultant whey from cheese industry is more than a million ton per a year. The whole quantity is considered as a waste, with the exception of a very little amount is used for animal feeding. Two types of whey are produced in Egypt, i.e., salted and unsalted. The salted one is mostly the predominant whey type, as it comes from the most popular cheese in Egypt which is called "Domiatti cheese". The present work was undertaken to determine the lipid materials of salted and unsalted whey as a by-product of cheese industry.

Methods: The lipid materials of freeze-dried whey were extracted using $CHCl_3$: MeOH(2:1, v/v), and saponified with methanolic KOH (20% v/v) for 24 hr. at room temp. The methyl easters of the fatty acids and unsaponifiables and standard compounds were analysed with a GCV Pye Unicam chromatograph equipped with dual flame ionization detector and dual channel recorder. Phospholipids were precipitated by acetone and fractionated using TLC technique.

Results and Conclusion: The general chemical analysis show that the salted whey contained lower amounts of lactose, proteins and lipids compared with unsalted whey. Whey lipids contained Palmitic and oleic acids as the abundant saturated and unsaturated acids, respectively.The presence of salt quantitatively altered the concentration of short-chain fatty acids due to its salting out phenomenon. A wide variety of hydrocarbons was found and C_{22},C_{23} constituted over 75% of the total hydrocarbons. The presence of salt in whey remarkably changed the hydrocarbon profile. The most predominant phospholipid was phosphatidyl choline followed by phosphatidyl inositol. Once more, the salt changed qualitatively and quantitatively the whey phospholipid pattern and led to precipitate some of the highly polar phospholipids with cheese during milk processing.

COMPARISON BETWEEN BIOLUMINESCENT AND CHEMILUMINESCENT METHODS FOR QUALITY CONTROL OF MILK

L.CECCHI, R.LODI, A.ZECCONI, M.SOMENZI

Centro Latte C.N.R.-Università di Milano - Via Celoria, 2 - 20133 MILANO ITALY

Foreword

Many analytical procedures based on luminescence- in both forms of bioluminescence (BL) and chemiluminescence (CL)- were proposed for some analysis. Bioluminescence was first applied mainly to determine contamination: it's well known that bioluminescence is based on light emitted by luciferin excited by ATP molecule extracted from cells of microorganisms. This seems unsuccessful in milk analysis owing to difficulties arising from the presence of both somatic and microbial cells. Other problems arise from the reagents themselves, which can easily deteriorate. In several hundred trials reproducible results were not always obtained. The bioluminescent method did give substantial results only in the measurement of somatic cells. The CL techniques were proposed in the food industry (for oils, potato chips, fried foods) both as a research tool in food spoilage and as a quality control measure to determine the extent or rate of spoilage before and after processing and storage. Milk powder and reconstituted milk, like heat-treated milk, were tested by different AA., who succeeded in recognizing them from raw milk. They used a "chemiluminescence induced" by light. We found that CL emitted by a milk sample is enhanced by a liquid scintillation cocktail, and the intensity of CL emission is dependent on the temperature and time of heating process. The intensity of this emission is reduced appreciably from raw to UHT milk. A simple measurement allows us to define the intensity of heat-treatment and consequently the quality of milk. In fact it's well known that drastic heat-treatment can modify the milk composition and alter its nutritional value. The amount of soluble whey protein (SP) is a valid parameter in evaluating this and it's strictly connected to heat-processes. We found that CL emission changes in the same way

as SP value. The measurement of SP by chemical methods takes at least 2 days, however the measurement of CL takes only 5 minutes.
beta-counter (e.g. Philiphs PW 4540 - ^{3}H preset window, coincidence "off") or luminometer (e.g. Packard PICOLITE 6500) are suitable instruments for this analysis.

Analytical procedure

0,5 ml of milk sample are well mixed with 10 ml of INSTAGEL (from Packard) and immediately put in the counting chamber of the beta-counter (in the authomated luminometer used only 0,1 ml of milk are mixed with 2 ml of INSTAGEL).
Counting data are collected after 1 min and recorded for the first 5 min. Results with these equipments are expressed in cpm.

Results

The data collected show:
- raw milk presents CL values higher than 200,000 cpm
- UHT milk presents CL values lower than 40,000 cpm
- in pasteurized milk CL values range from 120,000 to 60,000 cpm depending on the heat processes.

In the 70% of the trials CL values are strictly connected to SP values.
Samples of milk coming from the same dairy analyzed after 2-3 months gave identical CL values.
CL values are directly proportional to the amount of milk (at least in the range from 0,5 to 2 ml milk/10ml INSTAGEL).

Conclusions

By a simple and rapid method it seems possible to recognize a raw milk from a heat-processed one (pasteurized or UHT). By this analysis different pasteurization conditions can be evaluated allowing us to define the quality of milk. This can be done more rapidly than by SP determination.

RAPID DETERMINATION OF NITRATE AND NITRITE IN MILK AND MILK PRODUCTS BY CONTINUOUS FLOW INJECTION ANALYSIS

Lisbeth Munksgaard and Lars Thymark
The Danish Government Research Institute for Dairy Industry, Roskildevej 56, DK-3400 Hillerød, Denmark

Introduction

Analysis of nitrate and nitrite in milk and milk products is carried out in several laboratories for quality control and research laboratories. The IDF reference methods (1) are reliable, but very time consuming. So there is a need for routine methods which are reliable as the IDF reference method but much faster.

Materials and Methods

A procedure for automatic determination of nitrate and nitrite in milk and milk products was published earlier by Nijhuis et al. (2) for airsegmented continuous flow analysis. Continuous Flow Injection Analysis (FIA) is a new technique developed for automatic chemical analysis by Ruzicka & Hansen (3). The principle of this technique is injecting a liquid sample into a continuous, unsegmented carrier stream.Varying reagents are then added to the stream and allowed to react with the sample until the sample zone reaches the detector. The advantage of this technique to other automatic techniques (e.g. the airsegmented flow system) is that the result is obtained within a few minutes after injection of the sample, and the result can be obtained before the steady state level of the reaction has been reached. The theory for this is given in detail by Ruzicka & Hansen (3).
An instrument set up of the FIA system for determination of nitrate and nitrite in water is available (4). The principle is reduction of nitrate to nitrite by copperized cadmium and determination of nitrite is an unreduced and a reduced sample by detection of the intensity of a colour developed by the nitrite, sulphanilamide and hydrochloric N-1-napthyl-ethyldiamine. The same principle is used in the IDF reference methods for determination of nitrite and nitrate in milk products (1).
In our study we adopted the FIA method to analysis of milk

products. The main problems were to establish suitable procedures for removal of protein and fat from the samples and to make high viscous samples ready for introduction into the injection valve which has an inner diameter of only 0.5 mm.

The system is set up as illustrated in fig. 1 and 2. 200 µl of sample is injected into a carrier-stream of NH_4Cl buffer (pH 9.6). The sample is moved by the carrier into a dialysis cell where low molecular components (e.g. NO_2 and NO_3) pass through the membrane into a receptor stream of NH_4 Cl solution (pH 6) while the high molecular fraction goes to waste. For nitrate analysis the dialysate is carried on to the reductor, a 5 x 80 mm pipe filled with copperized granules. For nitrite analysis the reductor is bypassed. Then the stream is mixed with the colouring reagents and carried through the flow cell of a spectophotometer than continuously records the colour intensity at 538 mm.

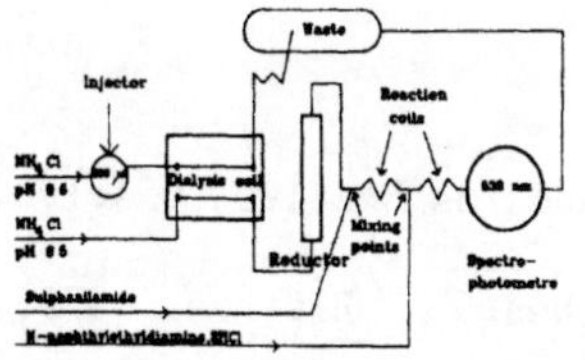

Fig.1 Flow sheet of the FIA-method

Fig.2 Photo of the instrumentation.

The analysis time i.e. the time from sample injection until detection is 1 minute for determination of NO_2^- and 1.5 minutes for NO_3^- analysis.
Samples of milk need no preparation before analysis, for samples of cultured milk we found pH should be increased to about 9 with a few drops of a concentrated NaOH-solution. This reduces the viscosity of the samples so they can be injected into the FIA-system. Samples of milk powder are to be reconstituted before injection, and we found that pH should be elevated by addition of NaOH as this improves the recovery of nitrate markedly. The improved recovery at higher pH may be due to the change of net charge of the protein. At higher pH the proteins are negatively charged so they cannot bind NO_3^- and NO_2^- as counter-ions. This may facilitate the dialysis of the negatively charged ions.

Results

The results obtained by FIA are on the same level as the results obtained with the IDF-reference methods (1), and the difference between 2 determinations on the same sample is reduced compared to the IDF-method.

References:

1. International IDF Standards 84A:1984,95A:1984and 97A: 1984.
2. Nijhuis, H., W. Heeschen & A. Blüthgen 1979. Milchwissenschaft 34 (7) 414-416.
3. Ruzicka, I. & E.H. Hansen 1981. Flow Injection Analysis.
4. Tecator Application Note AN 62/83.

DETERMINATION OF MILK PROTEINS BY IMMUNO-ELECTROPHORESIS

KR. ROTVIG KRISTIANSEN & A.-M. BECH
The Danish Goverment Research Institute for Dairy Industry, Roskildevej 56, DK-3400 Hillerød, Denmark

INTRODUCTION

Immuno-electrophoresis is a simple and specific technique for determination of individual proteins. By crossed immuno electrophoresis of whey with antisera to bovine whey 29 immuno precipitates were found (Løwenstein et al).

Therefore it was assumed that monospecific antisera to milk proteins might be used for quantative determinations of individual milk proteins. Whey proteins, however, are easily denatured by heat, which may destroy their immuno chemical characteristics. Caseins, on the other hand, are found to be very heat stable.

The purpose of our experiment was to investigate the suitability of immuno-electrophoresis for qualitative and quantatitive analysis of milk proteins in different types of milk and milk products. Crossed immuno-electrophoresis and rocket immuno-electrophoresis were carried out according to Axelsen et al.

In our laboratory immuno-electrophoresis has been used for several purposes :

1. Determination of whey proteins in raw and heat treated milk and whey samples.
2. Identification of peaks formed in crossed immuno-electrophoresis by the individual milk proteins.
3. Determination of rate constant, order of reaction and shift in reaction mechanism for the heat denaturation of whey proteins.
4. Demonstration of immunogenic identity between α_{s1}-casein and β-caseins.
5. Investigation of the suitability of rocket immuno-electrophoresis for determination of total casein in milk samples.

6. Investigation of the possibility of determining unhydrolysed α_{s1}-casein in cheeses.

TOTAL CASEIN IN MILK SAMPLES

The suitability was investigated of determining total casein in milk samples by rocket immuno-electrophoretic quantification of α_{s1} and β casein in raw milk and boiled milk samples. The best agreement with IDF International Standard No.29: 1964 results was obtained by determination of α_{s1}-casein with monospecific antisera.

Correlation between IDF-casein and α_{s1}-casein in raw milk was 0.78 and between α_{s1}-casein in raw and boiled milk was 0.96 Though, the standard deviation for the determination of a protein by rocket immuno-electrophoresis is 3-4% of the content, the mean square error for determining IDF-casein from α_{s1}-casein was close to 10%. This result is due to variations in casein composition in different milk samples.

CONCLUSION

Rocket immuno-electrophoresis with monospecific antisera was found suitable for quantitative determination of five individual milk proteins, with a 3-4% deviation. The whey proteins (α-lactalbumin, β lactoglobulin and serum albumin) could be determined in their native state only, not in denatured state. Heat treated α_{s1} and β casein reacted like unheated caseins and were easily determined.

By calculation of the total casein content in milk from the content of one casein (α_{s1}) the mean square error fo regression to IDF-casein was 10% of the casein content.

It was possible to determine the rate of denaturation and the order of this reaction for the 3 whey proteins. Quantitation of denatured whey proteins and of casein after some proteolysis was not possible due to atypical peaks.

ARRHENIUS PLOT

For the denaturation of β-lactoglobulin in skim milk. The bend is about 85° C indiates a change in the mechanism of reaction at that temperature.

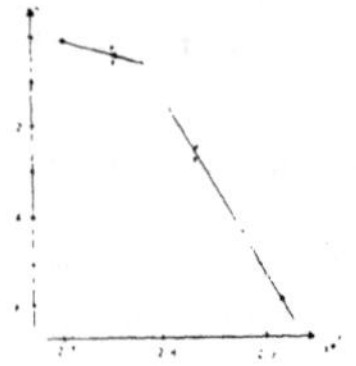

ROCKET IMMUNO-ELECTROPHORESIS

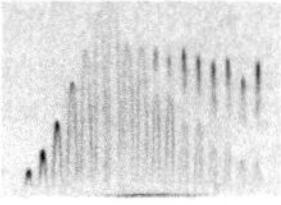

Determination of β-lactoglobulin in raw (peaks 8-9) and heat-treated (peaks 10-17) samples.

Identification of the α-lactalbumin (α-La) peak in crossed immuno-electrophoresis

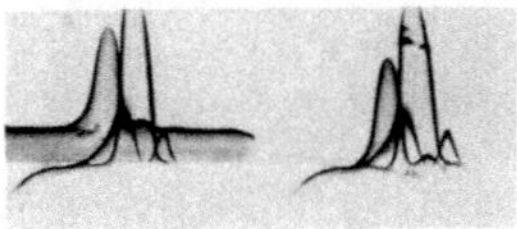

To the left 25 µg α-La in an intermediate gel results in a precipation line across the plate and the α-La peak to appears above the normal level. Notice the small peak to the left of and connected to the main α-La peak.

IMMUNOGENIC IDENTITY

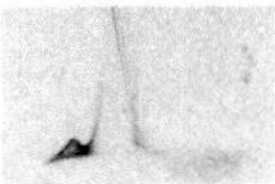

The large peak (β-casein) continues uninterrupted into the small peak to the left (γ-casein).

α_{s1}-CASEIN IN MILK AND FRESH AND RIPENED CHEESE

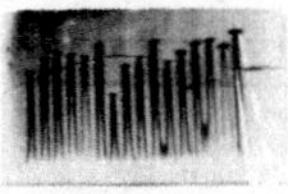

Peaks 7-10.6 and 16: varying amounts of refernce material and skim milk
Peaks 1-3 Ind 11-13: Cheese curd. 4 and 24 hour old cheese

Peaks 4 and 14, 5 and 15: 6 and 16 weeks old cheese.
Note.: These peaks appear indistinct due to proteolytic break down of α_{s1}-casein.

References:
Løwenstein H., Bjerrum O.J., Weeke E., and Weeke B. (1975).
Scand. J., Immunol. 4, suppl.2. 155-161.
Axelsen N.H., Krøll J. and Weeke B. (1973)
Scand. J., Immunol.2. suppl.1.

EFFECT OF CULTURE MEDIUM SUPPLEMENTS ON THE THIOCYANATE TOLERANCE OF STAPHYLOCOCCUS HYICUS MILK ISOLATES

J HARVEY* & A GILMOUR*[+]

*DEPARTMENT OF AGRICULTURE FOR NORTHERN IRELAND AND [+]THE QUEEN'S UNIVERSITY OF BELFAST, AGRICULTURE AND FOOD SCIENCE CENTRE, NEWFORGE LANE, BELFAST BT9 5PX, NORTHERN IRELAND

Introduction

In our laboratory we have been interested in the occurrence of *Staphylococcus hyicus* in bovine milk since this species has recently been shown to produce enterotoxin (Adesiyun 1981; Hoover *et al* 1983); the causative agent of staphylococcal food poisoning and hitherto more commonly associated with *Staph. aureus*. During studies on the isolation and identification of staphylococci in raw bovine milk (Harvey & Gilmour 1985) we observed that growth of the *Staph. hyicus* species was inhibited on the selective medium designed by Schleifer and Krämer (1980) for the isolation of all the known species of staphylococci. It was decided to investigate the cause of this inhibition with a view to improving the medium.

Materials and Methods

Test strains. (i) Fourteen different staphylococcal species isolated in our laboratory from raw bovine milk or obtained from a national culture collection.

(ii) Representatives of ten different genera commonly found in milk and obtained from national culture collections.

Media. (i) Schleifer and Krämer selective medium for staphylococci (SK) with and without supplements of 1% Tween 80 or 5% sheep blood or 5% egg yolk.

(ii) Non selective medium (NS) derived by removing the selective agents lithium chloride, glycine, sodium azide and potassium thiocyanate from SK.

(iii) NS medium with separate additions of the above selective agents at the concentrations used in SK.

Inoculation, incubation and examination of plates. Overnight broth cultures were inoculated on to the appropriate medium by one of two methods; either spread plating 0.1 ml aliquots of suitable dilutions or streaking undiluted broth cultures (in ten parallel streaks) on to the agar surface using a 1 µl loop. After incubation at 37° C for 24 h, counts were calculated from the spread plates or the extent of growth on the streaked plates was estimated by allocating a score of 1-10 depending on the number of streaks showing growth. Colony sizes were measured for both methods.

Results and Discussion

(i) *Staph. hyicus* subsp. *hyicus* and subsp. *chromogenes* produced pinpoint colonies or failed to grow on SK medium. All other staphylococcal species grew normally on SK.

(ii) Normal colony development of both subspecies of *Staph. hyicus* occurred on NS, NS + lithium chloride, NS + glycine and NS + sodium azide but not on NS + potassium thiocyanate. Therefore potassium thiocyanate was considered to be the cause of the growth inhibition of this species on SK.

(iii) The inhibitory effect of potassium thiocyanate in SK with respect to the subsp. *hyicus* but not *chromogenes* was found to be reduced by supplementation with 1% Tween 80 or 5% sheep blood or 5% egg yolk. Thus such additions enabled SK to be used satisfactorily for the isolation of *Staph. hyicus* subsp. *hyicus*.

(iv) Addition of 1% Tween 80 to SK medium did not reduce its selectivity with respect to non-staphylococcal organisms.

References. Adesiyun, A.A. 1981 Dissertation Abstracts International B42 2317. Harvey, J. & Gilmour, A. 1985 Journal of Applied Bacteriology 59 207-221. Hoover, D.G. *et al* 1983 Applied and Environmental Microbiology 46 649-660. Schleifer, K.H. & Krämer, E. 1980 Zentralblatt für Bakteriologie, Parasitenkunde, Infektionskrankheiten und Hygiene, Abt. I Originale C 1 270-280.

THE EFFECT OF CARBON DIOXIDE ON GROWTH AND ENZYME PRODUCTION BY PSEUDOMONAS FLUORESCENS B52

M.T. ROWE AND A. GILMOUR

Department of Agriculture for Northern Ireland and The Queen's University of Belfast, Agriculture and Food Science Centre, Newforge Lane, Belfast BT9 5PX, Northern Ireland.

Introduction

The addition of CO_2 has been advocated as a means of preserving the quality of refrigerated raw milk by inhibiting the psychrotropic microflora present. Psychrotrophs, particularly strains of Ps. fluorescens, have been implicated in the spoilage of milk and dairy products because of their ability to elaborate heat stable proteases and lipases which can survive pasteurisation and even UHT heat treatments (Cousin, 1982). King and Mabbitt (1982) have shown that CO_2 increases the duration of the lag phase but has a minimal effect on the log phase of growth of Ps. fluorescens strain NCDO 2085. There is however a lack of information on the effect of CO_2 on the ability of strains of Ps. fluorescens to produce extracellular enzymes. This poster describes the effects of CO_2 on the growth and protease and lipase production by Ps. fluorescens B52.

Materials and methods

Ps. fluorescens B52 was grown in a simulated milk medium at 7° C in a fermenter under the following conditions: pot size, 5 l; working volume, 3.5 l; impellor speed, 200 rpm; duration of incubation, 10 d; aeration, filtered air was passed through the headspace at 4 l/min. The sterile medium was flushed with filtered CO_2 gas until a concentration of approximately 30 m Moles/l was achieved. Thirty millimoles per litre is the highest recommended concentration of CO_2 which does not give rise to instability of milk protein (King and Mabbitt, 1982). The aeration and agitation rate was kept constant throughout this period. Once the target concentration had been reached the CO_2 supply was halted and the medium inoculated with Ps. fluorescens B52 which had been grown in nutrient broth at 25° C overnight. Ps. fluorescens B52 was chosen because it has been shown to be a prolific producer of protease and lipase (Rowe and Gilmour, 1982). Samples were removed at daily intervals and subjected to a viable count (Yeastrel Milk Agar, 25° C for 2 d), protease assay (Chism et al., 1979), lipase assay (Rowe and Gilmour, 1982) and a CO_2 measurement using a GLC.

Results and discussion

Figure 1. Growth of Ps. fluorescens B52 at 7° C in a simulated milk medium with added CO_2 showing changes in viable count (□), protease (○), and lipase (△), activities and CO_2 concentration (◇).

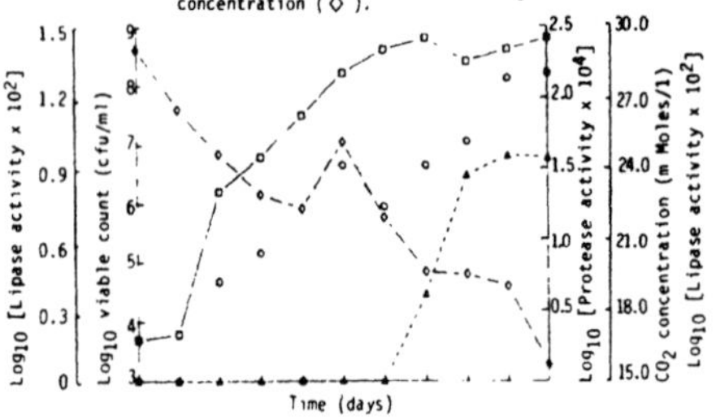

Figure 2. Growth of Ps. fluorescens B52 at 7° C in a simulated milk medium with added CO_2 showing changes in viable count (□), protease (○), and lipase (△), activities and CO_2 concentration (◇).

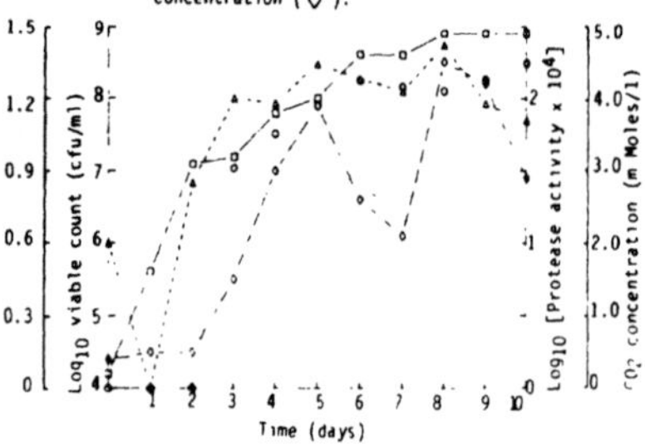

Carbon dioxide was observed to reduce the initial growth rate of the organsisms (Fig. 1) when compared to a control which had no added CO_2 (Fig. 2). Protease and lipase production were delayed by 2 and 5 days respectively (Figs. 1 and 2). Carbon dioxide was therefore shown to have an inhibitory effect on extracellular enzyme production by psychrotrophs perhaps by affecting the cytoplasmic membranes as proposed by Sears and Eisenberg (1961) which is the site of synthesis of extracellular proteins by bacteria (Fishman et al., 1980).

Work is in progress to repeat the above experiments using raw milk of varying quality, with its indigenous microflora, to determine the effect of CO_2 addition on growth and extracellular enzyme production by the psychrotrophs present.

References

Chism, G.W., Huang, A.E. and Marshall, J.A. (1979) J. Dairy Sci., 62, 1778.
Cousin, M.A. (1982) J. Food Protect., 45, 172.
Fishman, Y., Rottem, S. and Citri, N. (1980) J. Bact., 141, 1435.
King, J.S. and Mabbitt, L.A. (1982) J. Dairy Res., 49, 439.
Rowe, M.T. and Gilmour, A. (1982) Milchwissenschaft, 37, 597.
Sears, D.F. and Eisenberg, R.M. (1961) J. Gen. Physiol., 44, 869.

The incidence of Yersinia enterocolitica and Yersinia enterocolitica-like organisms in raw and pasteurized milk in Northern Ireland

S. J. Walker† and A. Gilmour*†

* Department of Agriculture for Northern Ireland and
† The Queen's University of Belfast, Agriculture and Food Science Centre, Newforge Lane. Belfast BT9 5PX, Northern Ireland

Introduction

Yersinia enterocolitica and Y. enterocolitica-like bacteria (Y. intermedia, Y. frederiksenii and Y. kristensenii) have been isolated from a wide range of foodstuffs throughout the world. Since certain serotypes of Y. enterocolitica are pathogenic for man, and may also grow at refrigeration temperatures, their presence in food is undesirable. Milk has acted as a vector of human disease transmission (Black et al. 1978).

Aim

To determine the incidence of Y. enterocolitica and Y. enterocolitica-like bacteria in raw and pasteurized milk in Northern Ireland.

Materials and Methods

Milk samples from the following sources were investigated:

(i) bulked raw milk (150 samples); (ii) farm bottled raw milk (20 samples),

(iii) farm pasteurized milk (50 samples); (iv) creamery pasteurized milk (100 samples).

Two enrichment procedures were investigated for each milk sample. Twenty-five ml of milk was added to:

(i) 225 ml of trypticase-soy broth (TSB) (for 24 h at 22° C), then 1 ml of this added to 9 ml a bile-oxalate-sorbose (BOS) selective medium (Schiemann 1982) (for 5 d at 22° C);

(ii) 225 ml of TSB and incubated for 21 d at 4° C.

Both enrichment cultures were plated on Yersinia Selective Agar (Difco), incubated at 25° C and examined after 24 and 48 h for colonies producing the typical 'bullseye' morphology. Presumptive isolates were confirmed using conventional biochemical techniques. All Y. enterocolitica isolates were biotyped (Bercovier et al. 1980) and serotyped (E. Fox, Yersinia Reference Facility, Leicester Royal Infirmary, Leicester).

Results and Discussion

Yersinia spp. were frequently isolated from both bulked raw milk and farm bottled milk (22.7% and 25.0% of samples respectively), with some samples containing two different species. Although not normally considered capable of surviving pasteurization, Y. enterocolitica was isolated from both farm and creamery pasteurized milks (8.0% and 6.0% of samples respectively). The presence of these strains may be attributable to post-pasteurization contamination.

Overall Y. enterocolitica was the species most commonly isolated, but Y. intermedia and Y. frederiksenii were also frequently obtained (52%, 31% and 15% of strains respectively). One atypical strain was identified as Y. aldovae.

Almost 75% of the Y. enterocolitica strains belonged to biotype 1, including some which could also ferment lactose, with the remaining strains from biotype 2. The major serotypes were 0:5,27, 0:34, 0:6,30 and 0 4, but one-third of strains could not be serotyped. The bio-serotypes obtained are not those commonly associated with human disease in Europe.

Enrichment of samples using the TSB-BOS regime allowed the recovery of 92.3% of all strains, and was considerably superior to cold enrichment in TSB which could recover only 15.4% of isolates. Neither method, however, could recover all the Yersinia strains.

References Bercovier, H. et al. 1980 Current Microbiology 4, 201-206; Black, R.E. et al. 1978 The New England Journal of Medicine 298, 76-79, Schiemann, D.A. 1982 Applied and Environmental Microbiology 43, 14-27.

RAPID DETECTION OF FLAVOUR DEFECTS IN MILK AND MILK PRODUCTS WITH AN AUTOMATIC SYSTEM FOR PURGE-AND-COLD-TRAPPING/CAPILLARY GAS CHROMATOGRAPHY

H T BADINGS, C DE JONG and R P M DOOPER[1]
Netherlands Institute for Dairy Research (NIZO), Ede, the Netherlands
[1] Chrompack International BV, Middelburg, the Netherlands

SUMMARY

A system for rapid automatic analysis of volatile compounds by purge-and-cold-trapping/capillary gas chromatography is described It is suitable for analysis of volatile compounds in a wide variety of samples, such as water, food products and environmental samples Possibilities and limitations of the PTI-system are evaluated in relation to several important parameters. The efficiency of different types of cryogenic trap (open tubular, coated, packed) is also investigated, it depends on purge-flow rate, temperature of trapping and total purge volume Examples are given of the analysis of volatile compounds in foods and water

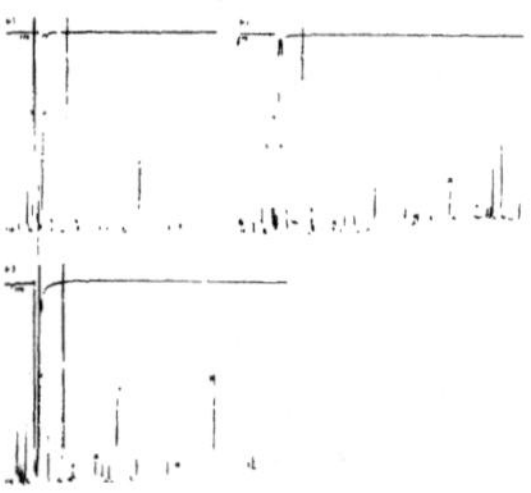

Fig 6 PTI/CGC analysis of volatile compounds from low-pasteurized milk (6-1), UHT-milk (6-2) and sterilized milk (6-3)
Identity of peaks, numbers refer to legends in Fig 5

Table 4. Results of PTI/CGC-analyses for a number of dairy products with different flavours.

product	flavour	H_2S	CH_3SH	CH_3CHO	CH_3SCH_3	Me butanal	2-pentanone	hexanal	CH_3SSCH_3	2-heptanone	heptanal
		volatile compounds (µg/kg)*									
LP-milk	normal	0 5	–	15	7	1	1	6	–	·	
LP-milk	malty		–		7	70	1	8	< 1	1	
LP-milk	unclean		–		55	1			–		·
LP-milk	spoilt		15		40	57					
LP-milk	"sunlight"		2	250	3	4			4		
UHT-milk	"UHT"	50	< 1		14	3	40		4	21	
sterilized milk	"sterilized"	–	3		20	1	120		4	82	
yoghurt	normal		–	> 1500	3	16	11	1	0 1	5	1
cultured buttermilk	normal		–		3	3	2	0 5	–	1	0 5
butterfat	normal		–		4	3	0 5	5	–	0 5	0 5
	ketonic		–		1	1	> 2000	10	–	> 2000	5
	oxidized		–		2	2		30	–		55
cheese (Gouda)	young (6 w)				4	26	12	2		10	
	old (1 year)				35	200	160	15		72	

Data of products other than milk are only indicative, because of lack of calibration curves for these products Given concentrations are examples, differences between samples of the same flavour quality may occur as a result of differences in starting material, processing, storage, etc

– = not detected · = not determined underlined figures are related to the (off)flavour mentioned

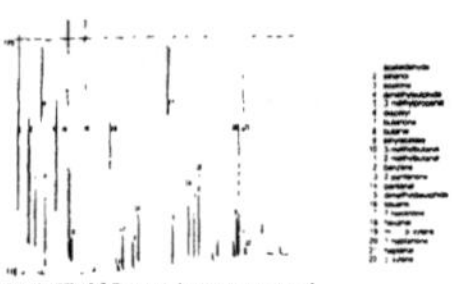

Fig 5 PTI/CGC run with a test mixture of volatile compounds in water

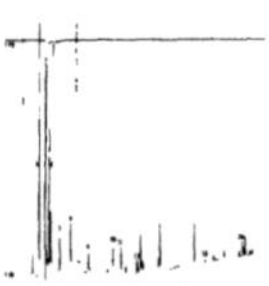

Fig 8 Analysis of volatile compounds from yoghurt For peak numbers see Fig 5

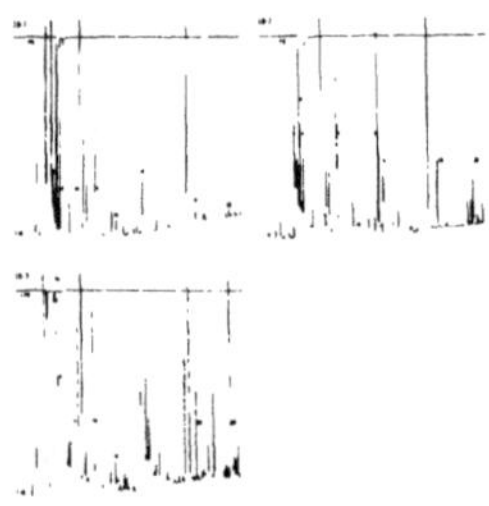

Fig 10 Analysis of volatile compounds from melted butterfat
10-1 fresh butterfat,
10-2 oxidized (fishy) butterfat,
10-3 mouldy butterfat
For peak numbers see Fig 5

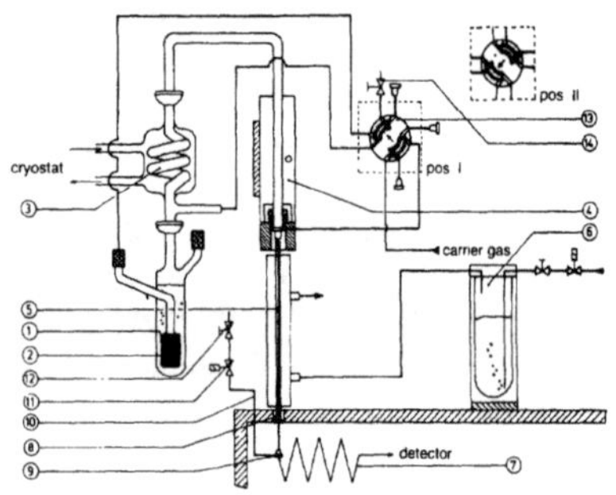

Fig. 2. Chrompack PTI system for purge-and-cold-trapping/capillary gas chromatography.

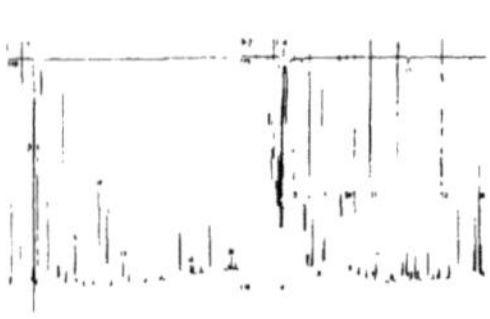

Fig 9 Analysis of volatile compounds from Gouda cheese
9-1 six week old cheese,
9-2 one year old cheese,
For peak numbers see Fig 5

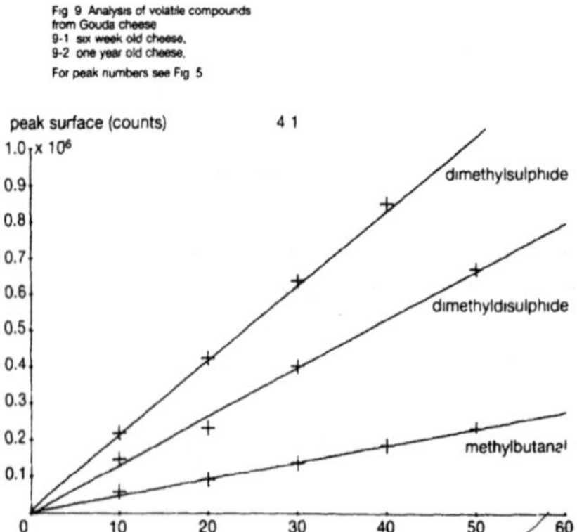

Fig. 4. Calibration curves of a number of volatile compounds in water.

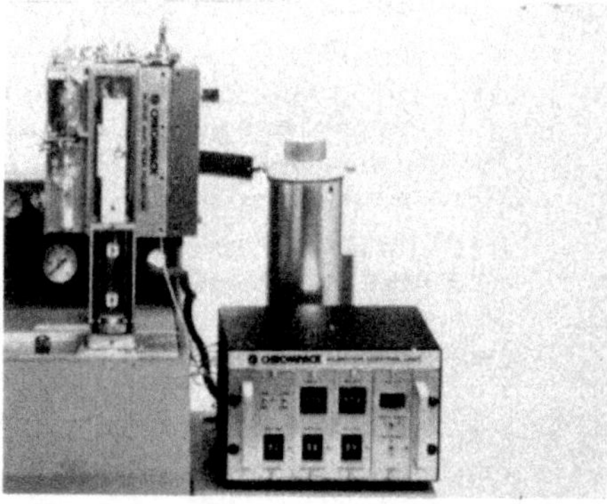

Lit. Ref.: J. High Resoln. Chromatography & C. C. 8(1985) 755.

The Using of Isotachophoretic Analysis in Dairy Analytic

Žabková J., Dairy Institute, Českobrodská 1174, Prague 198 00 - Czechoslovakia
Forman L., - " -
Pechačová, M., - " -
Obermaier O., - " -

In effort to extend milk and milk products analysis, the new method capillary isotachophoretic analysis (ITP) was used for the determination of organic acids in milk. This method enables to analyse native samples or only diluted by distilled water and unaffected by isolation technique. The application of volume-coupling arrangement allows the determination of macro- and microcomponents in one analysis with high acsuracy. ITP analyser of URVJT (Czechoslovakia) was used to analysis. Analysed samle was applied between two electrolytes (leading and terminating) and fractions were separated in high voltage eletric field in two separate capillary tubes from fluorinated ethylene-propylene copolymer (ID 0,8 and 0,3 mm) in volume-coupling arrangement. Separated components were determined by conductivity detector. The organic acids resulted from milk fermentation and C_1-C_{10} fatty acids resulted from milk fat lipolysis were determined. In comparison with ITP analysis the volatible acids distillation method was used, when only the total acidity of sample was obtained, without determination of individual acids distribution. According to our results, only about 6,5 % acids presented in original sample is obtained for titration. On the other hand the ITP method enables to determine both total sample acidity and the distribution of individual acids without complicated pretreatment of sample. This is evident advantage even in comparison with determination of lower fatty acids by ionex or silicagel chromatography, in which a complicated pretreatment of sample is necessary and accuracy is only 10-15 %. The ITP analysis is substantial quicker (20 min), more sensitive, more accurate (±1%) and more simple.

The problems concerning of different fatty acids solubility in water and organic solvents which complicates all analytical methods of milk fat lipolysis used to now, can be overcome by ITP method in water alcohol medium. The quantitative determination of C_1-C_{10} resp. C_1-C_{18} fatty acids is possible in one analysis with adequate water-methanol proportion.

TRANSFORMATIONS OF LACTOSE DURING THERMAL TREATMENTS OF SIMULATED MILK ULTRAFILTRATES.

A.Olano (*), N.Corzo (*), M.Paez (**) I.Martinez-Castro (**).

(*) Instituto de Fermentaciones Industriales (CSIC).
(**) Instituto de Química Orgánica (CSIC).
Juan de la Cierva, 3. 28006 Madrid.

Simulated milk ultrafiltrates (SMUF) were prepared according to Janness and Koops (1) in order to investigate the role of saline system of milk on isomerization of lactose during heat treatment.

Thermal treatments were carried out on two types of samples: liquid solutions and lyophylizates (Model systems for liquid and dried milks). No nitrogen was present in these solutions in order to avoid Maillard reactions. In some cases, variable amounts of lysine were added.

Lactulose, epilactose and galactose were formed in all samples, increasing with pH, temperature and time. A noticeable increase in lactose isomerization was observed when SMUF were prepared without calcium. This effect was observed in liquid solutions as well as in lyophylizates. The isomerization in lyophylizates was lower than in solutions.

Addition of lysine to SMUF without rising the pH did not cause a noticeable change on the lactose isomerization.

According to these results, it can be concluded that the saline milk system caused most of the lactulose formation during heat treatment of milk.

References

(1) R. Jenness and J. Koops. Neth Milk Dairy J., 16 (1962) 153-164

^{31}P NMR STUDY OF THE PHOSPHORYLATED COMPOUNDS IN MILK

M. Wåhlgren
Department of Food Technology, University of Lund,
P.O. Box 124, S-221 00 Lund, Sweden

^{31}P NMR has been used for the study of phosphorylated compounds in milk serum and milk ultrafiltrate. Resonances in the ^{31}P NMR spectra have been assigned to inorganic phosphate, N-acetylglucosamine-1-phosphate, glycerolphosphorylcholine, glycerolphosphorylethanolamine, phosphocreatine, galactose-1-phosphate, phosphorylcholine, phosphorylethanolamine, 3-phosphoglyceric acid, glycerol-1-phosphate and glucose-6-phosphate. We have also determined the concentration of these milk constituents.

PREDICTION OF SHELF LIFE OF UHT-PROCESSED MILK THROUGH VISCOSITY CHANGE

Dinesh K. Sharma and D.N. Prasad
Ultrastructure Laboratory, Dairy Microbiology Division,
National Dairy Research Institute, Karnal - 132001, India

In tropical countries, analysis and shelf life of the UHT processed milk are made by the bacteriological quality. However, the UHT - resistent enzymes which play an equally important role in the keeping quality are ignored. The processed milk, on storage, undergoes a variety of changes, viscosity being one of them, which is maximally effected. However, there is a paucity of any systematic information in respect of viscosity studies predicting shelf life. In the present study, Standardized (4.5% Fat, 8.5% SNF) and skimmed (0.1% Fat, 9.9% SNF) Buffalo milks were UHT - treated from 120°C - 150°C temperatures at interval for 2.8 seconds in an Indirect commercial UHT-system. Processed product was stored over a period of six months at ambient temperature. Viscosity changes were measured weekly using an Ostwald viscometer. A gradual change in the viscosity of the product was observed, leading to gelation. From the above observation an attempt has been made to formulate the shelf-life of UHT-milk. This formulation will be of great help, since viscosity changes due to minor alteration in bacterial, chemical or physical change in the product.

BUTTER MOISTURE - REFERENCE METHOD

M. PETERSEN, M. JOERIN*, E VIGGERS**
MOUNT MAUNGANUI REGIONAL DAIRY LABORATORY, MAF, P O BOX 5201, MOUNT MAUNGANUI, NEW ZEALAND.
*NATIONAL DAIRY LABORATORY, MAF, RUAKURA AGRICULTURAL RESEARCH CENTRE, PRIVATE BAG, HAMILTON, NEW ZEALAND.
**BIOMETRICS SECTION, MAF, PRIVATE BAG, WELLINGTON, NEW ZEALAND.

Precision data were determined from results of 12 laboratories analysing in duplicate 5 samples for butter moisture by a possible new reference method in New Zealand in July 1985. Sample moistures ranged from 15.2 to 17.1%.

The repeatability, r, (the value below which the absolute difference between two single test results, obtained by the same operator on identical samples on the same apparatus in a short interval of time, may be expected to be with a probability of 95%) was 0.08% moisture. The repeatability standard deviation S_r, was .03%.

The reproducibility, R, (the value below which the absolute difference between two single test results obtained by this method on identical material under different conditions may be expected to be with a probability of 95%), was 0.14%. The reproducibility standard deviation S_R was .05% moisture.

Overall analysis indicated that the capability of the method (the value of R obtainable if the method can be tightened to eliminate overall lab bias, but leaving the labx sample interaction unaltered) is 0.11.

These experimental results were obtained from laboratories mainly (6/12) unfamiliar with the new method.

*Changes proposed from the IDF method are:
*Changes in drying times and temperature.
*More critical control of initial sample taking and handling.
*More critical control of sample handling before weighings.
*Incorporation of Quality Control checks in the procedures.

DETERMINATION OF FAT AND TOTAL SOLIDS IN FLAVOURED MILK

K.L. ARORA & K.K. KALRA
National Dairy Research Institute,
KARNAL 132 001, HARYANA, INDIA

Gerber method for the estimation of fat content and the lactometric method for the estimation of total solids (TS) of flavoured milk were modified. For determining fat content by Gerber method, Gerber acid was replaced by sulphuric acid of sp. gravity 1.80 and 80 per cent concentration to avoid blackening of fat column due to the presence of cane sugar. For lactometric determination of TS, flavoured milk was diluted with distilled water in the ratio 1:1 by weight for reducing its sp. gravity to be measured by the lactometer. TS of the diluted sample was determined by both Quevenne and Zeal lactometer by using the formulae:

a) for Quevenne, % TS = 0.223 CLR + 1.18 F + 0.73
b) for Zeal , % TS = 0.225 CLR + 1.17 F + 1.00

where, F = % fat of the diluted sample
CLR = corrected lactometer reading at the respective temperature of 60°F for Quevenne and 84°F for Zeal lactometer

The values thus obtained were multiplied by 2 to get TS value of undiluted sample.

Ghee,the Indian counterpart of clarified butter fat, is always preferred due to its highly pleasing flavour and the contribution of free fatty acids(FFA) towards the flavour is well recognised.

A SIMPLE METHOD FOR THE PREPARATION OF METHYL ESTERS OF FREE FATTY ACIDS OF GHEE FOR ONWARD GLC ANALYSIS.

M.P.Bindal and D.K. Sharma
National Dairy Research Institute,KARNAL 132001, INDIA.

A simple method for the preparation of methyl esters of FFA without their prior isolation includes the incubation of ghee (0.1 g) with 14% BF_3 -methanol reagent (1 ml) in the presence of methyl urea dissolved in dichloro-methane) at $100^{o}C$ for 5 mts. The resultant methyl esters were washed with light petroleum(4 ml) in the presence of saturated solution of NaCl(4 ml) and were analysed by GLC over 10% DEGS.

The model mixture of triolein,tripalmitin,acid C_{14} and C_{18} gave only 2 peaks in the presence of methyl urea and 4 peaks in its absence indicating that glycerides in the presence of methyl urea did not form esters.

The total concentration of FFA(mg/g fat)in cow ghee (5.9-12.3) was 1.0-1.7 times that of buffalo ghee(5.8-7.5). The level of long chain (C_{16} and above)FFA in cow ghee was about double that of buffalo ghee whereas that of medium chain(C_{10} -C_{14}) in Cow (1.2-2.3) was slightly lower than in buffalo ghee(1.6-2.4). The short chain(C_4 and C_6)FFA were either absent or present only in trace amounts. Significant variations in FFA levels were also observed in ghee prepared by different methods, viz., Direct cream ghee (5.8-7.3), Creamery butter ghee (6.0-7.2)and Desi ghee (7.6-12.3). Highly significant variations in the level of C_{10}, C_{16}, C_{18} and $C_{18:1}$ acids may be responsible for flavour differences in cow and buffalo ghee.

PROTON-NMR-RELAXATION STUDY OF WATER AND FAT IN MILK PRODUCTS

B.A. Surkov, G.V. Murunova, I.P. Smirnov
All-Union Research Institute of Butter and Cheesemaking,
152620 Uglich, Krasnoarmeiskii Boulevard, 19, USSR

Spin-lattice (T1) and spin-spin (T2) relaxation times for protons of water and for aliphatic protons of fat in some milk products were measured at 25°C and resonance frequency 80.13 MHz. T1 and T2 for water protons were higher in butter and creams (20-83% fat) than those in cheeses (processed and natural) in accordance with more "bound" state of water in cheeses. In contrast, T1 and T2 for aliphatic protons of fat varied slightly both in cheeses and in butter and creams.

DETECTION OF COWS' MILK PROTEIN IN RIPENED COAT MILK CHEESE, USING ISOELECTRIC FOCUSING

E-L. Syväoja
Valio Finnish Co-operative Dairies' Association
Research and Development Department
P.O. Box 176
SF-00181 Helsinki
Finland

Polyacrylamide gel isoelectric focusing was used to detect the addition of cow's milk to goat's milk used to prepare fresh and ripened cheeses.

Samples were pasteurised goat's milk, cow's milk, and fresh and ripened cheeses made from mixed goat's milk containing known quantities of cow's milk. The caseins of the milk and cheese samples were purified and dissolved in 7 M urea containing dithiothreitol. The isoelectric focusing was carried out in polyacrylamide gel (5 % T and 3 % C) containing 7 M urea and 2 % ampholine pH 4,0 - 6,5 according to the method of Trieu-Cuot and Gripon.

A zone (pI 7.0) originating from cow's milk casein was absent from the pattern of goat's milk casein. This zone was separated with ampholine at pH 4 - 6.5 but not with it at pH 4-9 or 5-8. The lowest addition of cow's milk tested was 5 %, which resulted from the addition of starter culture prepared in cow's milk. The 5 % addition was clearly seen, so it is possible to detect much lower concentrations.

The proteolysis during cheese ripening was followed for 13 weeks and the electrophoretic patterns prepared. It was found that the pI 7.0 zone remained unchanged during the ripening of the cheeses.

The electrophoretic patterns were scanned with an LKB UltroScan Laser densitometer. The amount of indicator zone correlated closely with the proportion of added cow's milk casein.

Reference: Trieu-Cuot, P., and J.-C. Gripon 1981. J. Dairy Res. 48:303.

USE OF INOCULANTS AND ENZYMES AS GRASS SILAGE ADDITIVES

J. Setälä, T. Moisio and A. Rauramaa
Valio Finnish Co-operative Dairies' Association
Research and Development Department
P.O.Box 176
SF-00181 Helsinki
Finland

Direct-cut sward, about 3000 kg/treatment, was harvested with the flail harvester and ensiled in the bunker silo. The sward had dry matter (DM) content 12.6 % and 192 g crude protein and 40 g reducing sugars in kg DM. The preservatives were added in the harvester. When opening the silo, the silages were sampled after 231 days from the ensiling. The analytical procedures are described by PLAYNE and McDONALD (1966), STEINGASS (1983) and SETÄLÄ et. al. (1985).

Treatment	DM-%	pH	Lactic	Acetic	Propionic	Butyric	Sugars	NH_3-N, % in total N
			g/kg DM					
a) No additive	14.8	4.74	24.7	76.4	11.5	0.6	8.1	14.4
b) AIV II	18.5	3.77	50.5	15.6	0.7	0.8	17.1	4.2
c) Inoculant L	16.4	4.78	18.2	75.4	8.5	0.8	8.0	16.0
d) Inoculant LP	17.1	4.74	21.5	69.8	8.2	1.4	9.0	15.8
e) Cellulase	18.1	4.51	43.3	72.0	11.3	1.0	9.8	14.9

L = Lact. plantarum, 11×10^{11} cfu/ton;
LP = Lact. plantarum, Pediococcus acidilacti 8×10^{11} cfu/ton;
Cellulase = Trichoderma viride, 1.8×10^{8}IU/ton; AIV II, 5 liters/ton

The best quality silage was obtained with the AIV II acid mixture (80 % formic, 2 % orthophosphoric acid). Both the protein value and the energy values (organic) matter digestibility in vitro, OMD-%) of the silages were decreased when the other ensiling method than the acid treatment was used. The content of the amines (= cadaverine, putrescine, tryptamine, tyramine, β-fenyletylamine) in the silages a, b, c, d and e was 9.2, 3.3, 11.1, 7.5 and 4.8 g/kg DM, and the OMD-% 65.5, 71.2, 67.1, 64.8 and 64.0, respectively.

References:

PLAYNE, M.J. & McDONALD, P. 1966. J.Sci. Food Agric. 17:264-268.
STEINGASS, H. 1983. Thesis, Universität Hohenheim, 188 p.
SETÄLÄ, J., TESFA, A., RAURAMAA, A. & POUTIAINEN, E. 1985. J.agric. Sci., Finl. 57: 139-146.

EFFECT OF HEAT TREATMENT ON QUALITY AND SHELF-LIFE OF YOGHURT

F.O. Mohammed, S.D. Al-Sawaf, M.T. Darkazly

Yoghurt manufactured by the conventional method in Iraq is prone to the common defects. Its heat treatment at 70 . 75 and 80°C for 5 minutes resulted in improving the quality and shelf-life upon 21 days of cold storage. Over 80 percent of the heat treated samples were still accepted and highly scored, while all of the untreated samples (control) were rejected at the end of storage. A slight change in both pH values and colony counts of heat treated yoghurt was observed.

Antimicrobial effect of zabady on yeast and coliforms.

S.A.Abou-Donia, I.M.El-Said, S.E.Rakshy, I.Sirry
Department of Agricultural Industries,
Faculty of Agriculture, University of Alexanaria
Alexandria, Egypt.

The mixed and single zabady starter (*Streptococcus thermophilus* + *Lactobacillus bulgaricus*) slightly decreased the growth of yeast. When zabady starter and *Escherishia coli* cultivated at 42°C, it was observed that the count of *E.coli* at zero time was 3.8×10^{6} cfu/ml., whent *S.thermophilus* was cultivated together with *E.coli*, there were no obvious changes in numbers of either *S.thermophilus* or *E.coli*, also when *L.bulgaricus* was cultivated together with *E.coli*, there were no obvious changes in number of either *L.bulgaricus* and *E.coli*.
To detect the inhibitory substances that may be produced by *L.bulgaricus* against *E.coli*, the conventional diffusion technique was used. A clear zon about 3 mm in width was observed surrouinding the glass cylinder.

LACTOBACILLUS ACIDOPHILUS IN THE GASTROENTERIC TRACT OF *S.FAECIUM*-TREATED GNOTOXENIC ANIMALS

P. CAMASCHELLA, B. BIANCHI-SALVADORI
Centro Sperimentale del Latte, Milan

INTRODUCTION

Streptococcus faecium and *Lactobacillus acidophilus* are used as eubiotic-bacteria in human and animal nutrition.

We observed "in vitro" the proprieties of these microorganisms and the enzymatic activities, such as acidification, lipolysis and proteolysis by both bacteria ad well as the amylolytic activity of *L.acidophilus* on the components of the food bolus.

The two species of bacteria are used as integrating principles in the diet: *S.faecium* is preferably given during the suckling period and *L.acidophilus* during weaning. Experiences of administration of *S.faecium* and *L.acidophilus* during these two feeding periods have resulted in improved health conditions in farm animals. Inoculated into germ-free mice, *S.faecium*, adheres promptly to the mucosal cells of the small and large intestine; in the small gut, where enteritis causing germs are likely to settle, it also proliferates, thus forming a barrier effect.

This work investigates the degree of colonization, in germ-free MICE, of *S.faecium* and *L.acidophilus* on the base of results obtained by studies of their physiological characteristics and of breeding animals.

MATERIAL AND METHODS

10 germ-free male mice inoculated with:
1.5×10^6 *Streptococcus faecium* cells
One week later with:
5×10^6 *Lactobacillus acidophilus* cells
1.3×10^6 *Bacillus subtilis thermophilus* spores (marker)

Microbial count:
S.faecium on MRS Agar
L.acidophilus on MRS Agar
B.subtilis thermophilus on Plate Count Agar at 60°C

Enumeration was carried out on:
faeces

small intestine } walls
large intestine } luminal contents

Tracts of gut were prepared for scanning electr on microscopy

by **critical point drying** { to avoid modification of the microbial and mucosal cells and image distortion.

Scanning electron microscopy

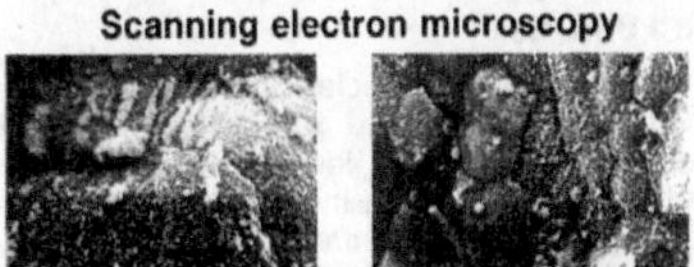

Small intestine

Large intestine

RESULTS

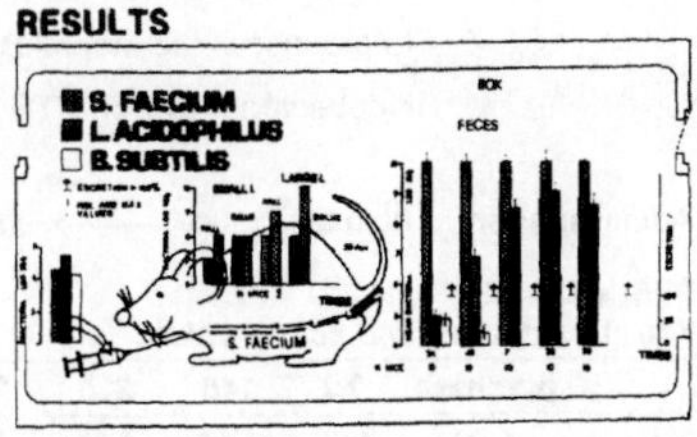

CONCLUSIONS

L.acidophilus gradually colonizes the gut of (*S.faecium*-treated) gnotoxenic mice within the first 48 hours and proliferates up to levels of 10^8 after 7 days. The lactobacillus adheres to the gut wall and its presence on the small intestinal wall indicates peristalsis-resistance with possible barrier effect against pathogen germs.

L.acidophilus may BE USED during the weaning period of animals treated with *S.faecium* from birth, to ensure the gut flora equilibrium.

VIABILITY DURATION OF THE MICROBIAL FLORA IN COMMERCIAL YOGURT-MIXES WITH DIFFERENT LACTIC ACID BACTERIA SPECIES

B. BIANCHI-SALVADORI, P. CAMASCHELLA
Centro Sperimentale del Latte, Milan

INTRODUCTION

Some dairy industries launch on the market new types of products consisting of different mixes of traditional fermented milks and cultures of bacteria belonging to the human intestinal flora, attracting the consumers' attention to products called therapeutic and/or dietetic.

In this investigation we examined a commercial mix product prepared with **yogurt, Lactobacillus acidophilus** and **Bifidobacteria**, in order to verify survival of these bacteria during storage at 4°C.

MATERIAL AND METHODS

4 series of fermented milk preparation from different shops, at different times.

Determination carried out:

pH

Titrable acidity

Microbical counts of LACTIC ACID BACTERIA (LAB):

Lactobacillus bulgaricus on MRS Agar at pH 5.4 }
Lactobacillus acidophilus on MRS Agar at pH 6,8 } cultural tests, biochemical tests, lactic acid isomers

Streptococcus thermophilus on M17 Agar — cultural tests, biochemical tests

Bifidobacteria (BF) on PTG broth + NPNL solution — morphological observation through a phase-contrast microscope (magnification 100 × oil immersion)

Administration to holoxenic mice ⟶ sampling of faeces ⟶ Bifidobacteria detection.

RESULTS

Tab. 1 *Mean pH and acidity values*

	purchase	7d	14d	21d	28d
pH	4.20	4.20	4.15	4.13	4.00
acidity	1.15	1.15	1.15	1.17	1.29

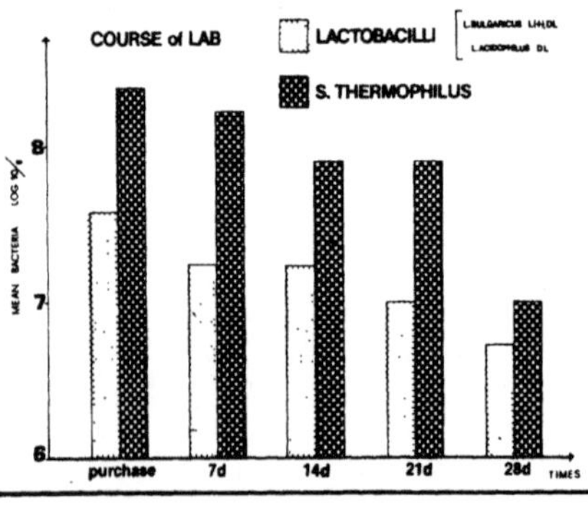

Tab. 2 *Course of Bifidobacteria*

	purchase	7d	14d	21d	28d
n°BFbacteria	present in 5 g (1 S)	0*	0*	0*	0*
	1.000/g (1 S)				

S = n° samples
* in 5 g

CONCLUSIONS

The various bacterial species present in the commercial product prepared with yogurt to which bacteria of the human digestive tube had been added, do not maintain the initial equilibrium during the storage period. This situation of incompatibility between the various species leads to a product that cannot be considered satisfactory to the consumer, because it does not supply the dietetic performace announced on the label.

Process alterations in shrikhand technology

R.S.Patel and B.K.Chakraborty
National Dairy Research Institute, Karnal
132001, India.

The conventional shrikhand making technology requires a relatively long time because of slow setting of milk through lactic fermentation. Moreover, this product being a summer speciality, its manufacture from fresh milk in the summer implies diversion of milk from fluid consumption during the lean season. Also, the shelf-life of the product is limited. Hence, a study on making certain process alterations in shrikhand technology was carried out. The results are summarized here.

Various cultures of lactic acid bacteria consisting of streptococci and lactobacilli were examined with respect to time required for curd formation and properties of the product. Yoghurt culture was the most desirable as it produced satisfactory curd within 4 hrs, thus reducing the total processing time from 8 to 12 hrs to 4 hrs for curd preparation. A standard method was established for preparing shrikhand samples. The method involved using buffalo skimmilk (10% total solids), subjected to heat treatment of 85°C/30 min, followed by 1.5 to 2.0% inoculation of an active yoghurt culture at 42°C for obtaining a curd (0.9% LA) at the end of 4 hrs. The chakka obtained from this curd was then blended with required quantity of 80% fat pasteurized (85°C/10 min) plastic cream and sugar to represent a final composition of 40% moisture, 6% fat and 41% sugar.

As an alternate source of milk solids, reconstituted skimmilk could successfully replace skimmilk for shrikhand making. Potassium sorbate at 0.05% level substantially enhanced the storage stability of shrikhand.

" CO_2 PRODUCTION IN KEFIR BY TORULOPSIS SPHAERICA IMMOBILIZED CELLS "

M.GOBBETTI, J.ROSSI, F.CLEMENTI
Institute of Dairy Microbiology, Faculty of Agriculture, University of Perugia, Italy.

Immobilized cell bioreactor used to incorporate CO_2 in fermented milk

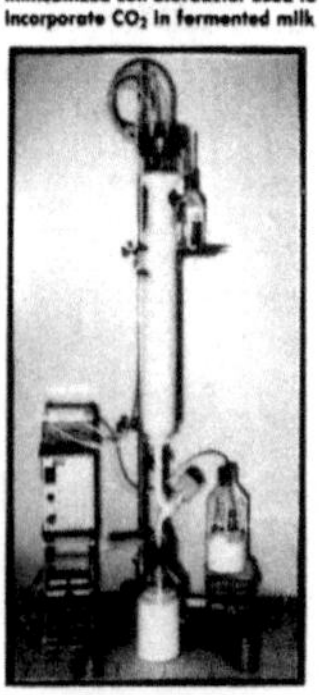

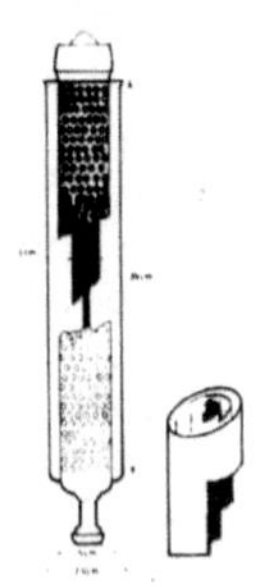

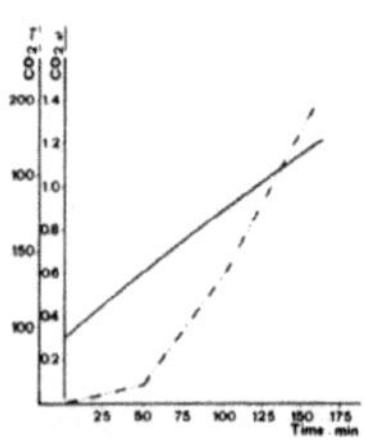

Continuous CO_2 production (— dissolved CO_2; – – gaseous CO_2), under standard conditions (2.7% Na-alginate; 13.7% -wet weight- cell concentration in the gel, flow rate 18.8 ml min⁻¹, 28°C)

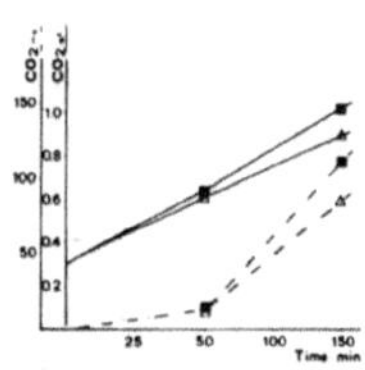

Continuous CO_2 production (— dissolved CO_2; – –gaseous CO_2) at different cell concentration in the gel (△△13.7%, ■■23.6%)

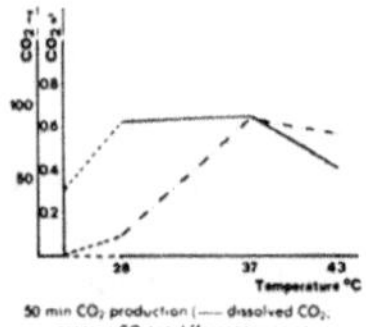

50 min CO_2 production (— dissolved CO_2; – –gaseous CO_2) at different temperatures

SEM micrographs of Ca-alginate immobilized cells (13.7%)

Gel bead, diameter 6 mm (20 x)

Bead section, before use (2500 x) CFU = 27.0 x 10^8/g

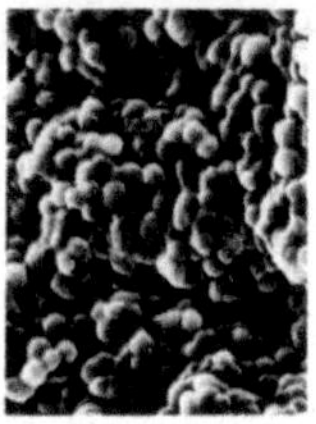

Bead section, after 15 h (2500 x) CFU = 85.0 x 10^7/g

Chemical and Microbiological features of the fermented milk

	Before treatmet	After treatmet 30 min	After treatmet 2 h 30 min
pH	4.1		
°SH %	30.0		
Protein %	3.9		
Ethanol g/l	traces	0.54	1.02
Acetaldehyde	traces	traces	traces
Dissolved CO_2 g/l (determined by enzymatic method)	0.30	0.61	1.15
Lactobacillus CFU/ml	14.5 x 10^7		
Streptococcus CFU/ml	19.3 x 10^7		

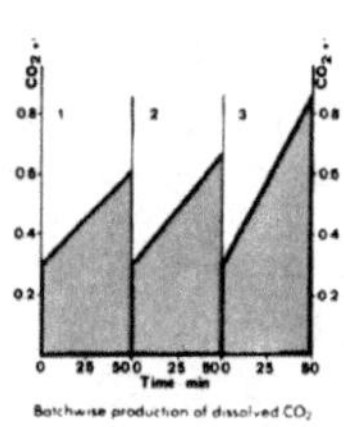

Batchwise production of dissolved CO_2

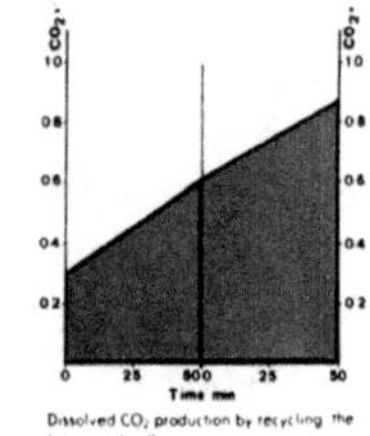

Dissolved CO_2 production by recycling the fermented milk

COMPARATIVE STUDY OF POLYSACCHARIDES OBTAINED FROM KEFIR GRAINS AND PRODUCED BY A HOMOFERMENTATIVE LACTOBACILLUS SPECIES AND LACTOBACILLUS KEFIR ISOLATED FROM KEFIR GRAINS

T.TOBA, K.ARIHARA* AND S.ADACHI
LABORATORY OF ANIMAL PRODUCTS TECHNOLOGY, COLLEGE OF AGRICULTURE, TOHOKU UNIVERSITY, TSUTSUMIDORI-AMAMIYACHO 1-1, SENDAI 980, JAPAN.
*THE SCHOOL OF VETERINARY MEDICINE AND ANIMAL SCIENCES, KITASATO UNIVERSITY, SANBONGI-MAEYACHI 149-2, TOWADA 034, JAPAN.

INTRODUCTION

In the previous paper(1), a polysaccharide-producing homofermentative Lactobacillus species K_1 has been isolated on the new medium(referred to as KPL) from kefir grains. The present study concerned with the chemical characteristics of the polysaccharides obtained from kefir grains, Lactobacillus K_1 and L.kefir.

MATERIALS AND METHODS

Kefir grains were purchased from Chr.Hansen's Laboratory (Copenhagen, Denmark). Lactobacillus K_1 and L.kefir were isolated from grains on KPL agar(1) and MRS agar, respectively. The polysaccharides were obtained from kefir grains, Lactobacillus K_1 and L.kefir by hot water (-HW) and acidic buffer (-A) extraction. The polysaccharides were also obtained their media(-M). Each polysaccharides were purified by ethanol precipitation, ion-exchange chromatography and gel filtration chromatography.

RESULTS AND DISCUSSION

Gel filtration chromatography showed K_1-M, K_1-HW and K_1-A had similar molecular weight with KG-M, KG-HW and KG-A ,respectively. KL had different molecular weight with the others. Sugar composition of the polysaccharides obtained from Lactobacillus K_1 was similar to those from kefir grains and different from L.kefir . Methylation analysis showed K_1-M, K_1-HW, K_1-A, KG-M, KG-HW and KG-A had similar chemical structures. It also indicated that the chemical structure of KL differ from the others. Thus we considered the polysaccharides produced by Lactobacillus K_1 were the same as those in kefir grains.

CONCLUSION

We considered Lactobacillus K_1 is responsible for polysaccharide production in the grains.

REFERENCES

(1) T.Toba, S.Abe, K.Arihara and S.Adachi: Agric.Biol.Chem., submitted for publication.

OPTIMIZATION OF THE CULTIVATION MEDIUM COMPOSITION FOR LACTIC ACID BACTERIA

Z. Libudzisz, B. Mansfeld, E. Kacki*, H. Oberman
Institute of Fermentation Technology and Microbiology,
*Institute of Computer Science, Technical University,
90-924 Łódź, B. Stefanowskiego 4/10, Poland

Using the experimental version of Gauss-Seidel's iterative method the composition of the medium has been established for optimal production of biomass and of optimal growth rate of mesophilic strains of lactic streptococci of the species: S. lactis, S. lactis ssp. diacetylactis, S. cremoris and Leuc. cremoris.

INFLUENCE OF β-LACTOGLOBULIN DENATURATION ON SYNERESIS AND RHEOLOGICAL PROPERTIES OF SET-STYLE NONFAT YOGHURT.

F. DANNENBERG AND H.G. KESSLER

Institute for Dairy Science and Food Process Engineering,
Technical University of Munich
D-8050 Freising-Weihenstephan

INFLUENCE OF β-LACTOGLOBULIN DENATURATION ON SYNERESIS AND RHEOLOGICAL PROPERTIES OF SET-STYLE NONFAT YOGHURT.

F. DANNENBERG AND H.G. KESSLER

INTRODUCTION

TEXTURE OF SET-STYLE YOGHURT IS GREALTLY INFLUENCED BY THE DEGREE OF INTERACTION OF CASEIN AND WHEY PROTEINS FOLLOWING DENATURATION. IN ORDER TO ASSES THE INFLUENCE OF THE DENATURATION OF β-LACTOGLOBULIN YOGHURT WAS PREPARED FROM SKIM MILK HEATED OVER A WIDE TEMPERATURE/TIME RANGE.

INVESTIGATIONS WERE BASED ON PREVIOUS REACTION KINETIC STUDIES ON WHEY PROTEIN DENATURATION. RESULTS CONCERNING β-LG B DENATURATION ARE SHOWN BELOW.

MATERIAL AND METHODS

HEATING WAS CARRIED OUT ON A PILOT HEATING PLANT AT TEMPERATURES BETWEEN 70 AND 130°C RESULTING IN WELL DEFINED DENATURATION RATES OF β-LG B BETWEEN 10 AND 99%. YOGHURT WAS PREPARED FROM THIS MILK UNDER STRICTLY CONTROLLED FERMENTATION CONDITIONS. SYNERESIS, FIRMNESS AND RHEOLOGICAL PROPERTIES OF THE COAGULUM WERE DETERMINED.

DENATURATION OF β-LACTOGLOBULIN B AS A FUNCTION OF HEATING TIME AND HOLDING TEMPERATURE.

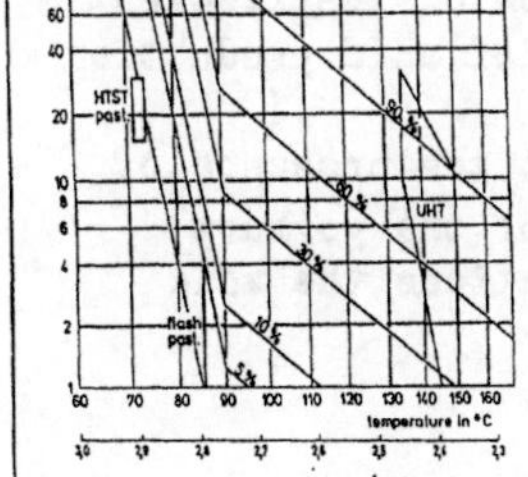

RESULTS

INFLUENCE OF β-LACTOGLOBULIN B DENATURATION ON PROPERTIES OF SET-STYLE YOGHURT MANUFACTURED FROM SKIM MILK THAT WAS PREHEATED DIFFERENTLY.

DRAINED WHEY (SYNERESIS)

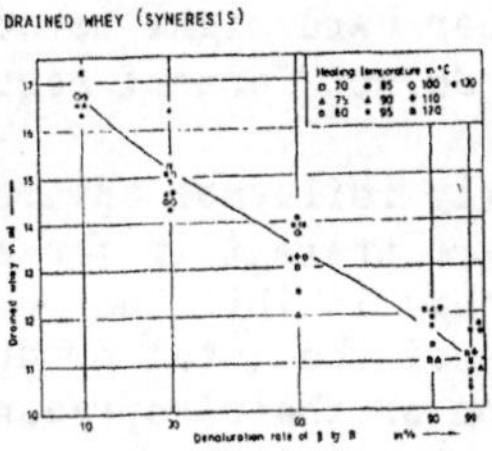

CURD TENSION (FIRMNESS)

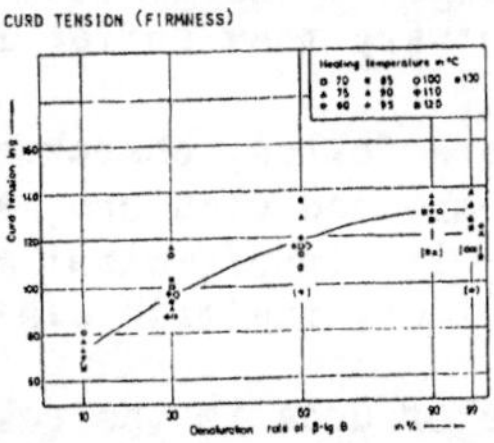

AREA OF HYSTERESIS LOOP (ASSESMENT OF THIXOTROPY)

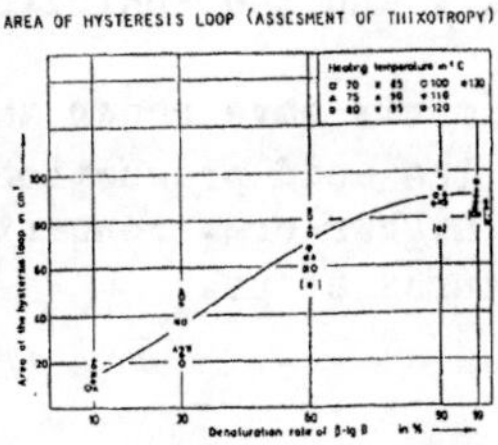

LINES OF CONSTANT DENATURATION RATE OF β-LACTOGLOBULIN B AND MEASURED VALUES OF DRAINED WHEY AS A FUNCTION OF HEATING TEMPERATURE AND HOLDING TIME.

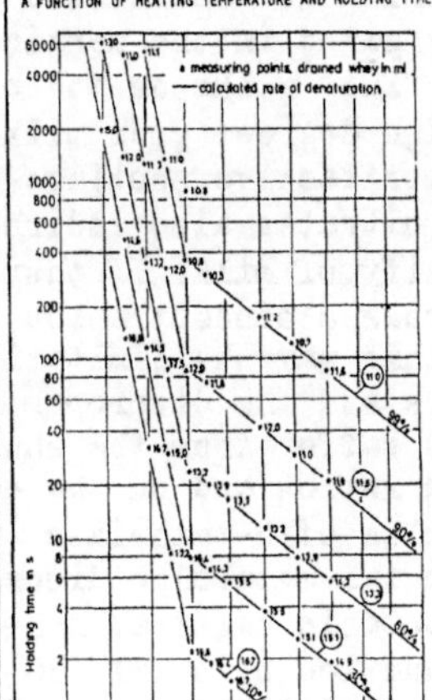

SUMMARY

THE EFFECT OF WHEY PROTEIN DENATURATION ON THE STRUCTURE OF YOGHURT GELS WAS EXAMINED. THE RESULTS SHOW THAT PROPERTIES OF SET-STYLE NONFAT YOGHURT ARE RELATED TO β-LG DENATURATION AND ARE LARGELY INDEPENDENT OF HEATING TEMPERATURE. INCREASING THE HOLDING TIME ABOVE THAT REQUIRED FOR 99% β-LG DENATURATION RESULTS NEITHER IN REDUCED SYNERESIS NOR IN IMPROVEMENT OF RHEOLOGICAL PROPERTIES.

INSTITUTE FOR DAIRY SCIENCE
AND FOOD PROCESS ENGINEERING
TECHNICAL UNIVERSITY OF MUNICH
D-8050 FREISING-WEIHENSTEPHAN

The Influence of Nitrates on Activity of Lactic Acid Bacteria

Cvak Z. - Podhorský M. - Hurťáková J.
Dairy Research Institute, Prague

The lactate-hydrogenase, alcohol-dehydrogenase as well as oxidases of cytochrome type are inhibited by the nitrates. If the nitrates are present in milk in concentrations higher than 5 mg . l^{-1} they are retarding the development and the fermentation activity of lactic acid bacteria.

The nitrate inhibition of lactic acid bacteria depends on the concentration relations. The fermented milk products, the processing of which is condicioned by the sufficient big innoculum, are not endangered in the mentioned respect.

The fluid milks on the other hand might be adversely influenced to a high degree; such milks do not ferment regularly and undergo a putrifactive spoilage.

The nitrates also indirectly influence adversely the fermentation ability of milk. If they are present in the cow nutrition in an increased concentration, they inhibit the ruminent fermentation. In this way the limitation of the total production of volatile fatty acids and the disproportion of their representation occurs. The milk cows suffer from the shortage of energy even with the well composed feed ration and on the contrary they suffer from the relative surplus of N-materials.

Then the metabolic disorders follow, charakterized by the impair of acidobasic balance, non-compensed alcalose, ketose reduction of milk production by 5 till 30 %, its insufficient acidity and the low fermentation capacity. Besides, the milk has a high content of urea: 2 - 6 nmol . l^{-1}.

Most probably the presence of urea is the reason of insufficient fermentation ability of milk and the inhibitor of acid production of lactic acid bacteria.

The number of model experiments have shown that the presence of 2 nmol . l^{-1} urea reduces the acid production of the yoghurt culture about by 50 %. At higher urea concentrations the acid production of yoghurt cultures stops.

Die Gewinnung von Mutanten von Milchsäurebakterien mit einer erhöhten Resistenz gegenüber Hemmstoffen und mit erhöhter antimikrobieller Wirksamkeit auf die unerwünschte gastrointestinelle Mikroflora.

Hylmar, H., Forschungsanstalt der Molkereiindustrie Prag,
Pokorná, L., - " -
Šalaková, A., - " -

Durch Selektion und Züchtung von Milchsäurebakterienkulturen, die zur Herstellung von Sauermilcherzeugnissen benutzt werden, wurden Stämme und Mutanten mit einer erhöhten Resistenz gegenüber Antibiotika und anderen chemotherapeutischen Stoffen gewonnen. Das betrifft vor allem solche Mittel, die in der Human- und Veterinärmedizin angewendet werden. Die Mutanten wiesen ausserdem eine erhöhte antimikrobielle Wirkung auf die unerwünschte gastrointestinelle Mikroflora und auf einige, die Milch kontaminierende Mikroben, auf. Die Aufmerksamkeit war vor allem auf eine vorteilhafte Kombination beider Eigenschaften gerichtet, damit diese ausgesuchten Stämme bei der Herstellung von Sauermilchprodukten mit diätisch-heilenden Eigenschaften auch als zusätzliche Behandlung für Patienten, bei denen Antibiotika zur Anwendung kamen, genutzt werden können.
Das umfassendste Spektrum antimikrobieller Wirkung wurde bei Stämmen von Lactobacillus acidophilus gefunden. Diese inhibierten z.B. das Wachstum von Stämmen von Pseudomonas aeruginosa, Pseudomonas fluorescens. Staphylococcus aureus, Staphylococcus agalactiae, Serratia marcescens, Escherichia coli, Klebsiella aerogenes, Microc. varians, Proteus vulgaris, Bacillus subtilis und Bacillus cereus. Aber auch Stämme von Lb. bulgaricus und Bifidobacterium wiesen eine grössere antimikrobielle Wirkung als Streptoccocus-Stämme der Milchsäuregärung auf. Für die Kultur Lb. acidophilus wurde nachgewiesen, dass neben der produzierten Milchsäure, gegebenenfalls den anderen organischen Säuren, und weiteren Stoffwechselprodukten, vor allem die gebildeten Antibiotika für die antimikrobielle Wirkung verantwortlich sind.
Aus den gewonnenen mutanten Stämmen der Milchsäurebakterien wurden schrittweise Mischkulturen mit neuen Eigenschaften zusammengestellt, oder sie wurden als Monokulturen zur Herstellung von fermentierten Milchprodukten mit erhöhter diätisch-heilender Wirkung benutzt.

LIVE LACTOBACILLI CONTAINING YOGURT IS EFFECTIVE IN STIMULATING HOST'S IMMUNOCOMPETENCE.

B. BIANCHI SALVADORI & C. DE SIMONE [§]
Centro Sperimentale del Latte, Milano, Italia and [§]Clinica Malattie Infettive, Università "La Sapienza", Roma, Italia

The addition of yogurt containing live strains of lactobacilli to the standard diet results, after two weeks of treatment, in a different expression on the murine splenocytes, as compared to the group of the control animals, of the surface antigens Th1,2, Lyt1, Lyt2. From a functional point of view, the proliferative responses of the splenocytes to the phytohemagglutinin (PHA), concanavalin A (Con A) and pokeweed mitogens (PWM) were also increased significantly. A brief treatment of the yogurt with heat deprives it of such properties. Experiments done employing lymphocytes isolated from mouse Peyer's Patches have shown that the ingestion of yogurt containing live lactobacilli as a supplement to the standard diet induces a significant increase in the mitogenic responses of the murine B lymphocytes to <u>E. coli</u> lipopolysaccharide (LPS) as compared to the control group fed a diet supplemented with milk in isocaloric quantities with those of the yogurt.

Surface markers (% splenocytes)

		Thy 1, 2	Lyt 1	Lyt 2	sIg
15 days	Group A	35 + 8	40 + 7	20 + 2	23 + 3
	Group B	48 + 8*	35 + 4	15 + 4	23 + 4
	Group C	55 + 8*	44 + 6	17 + 3	23 + 9
	Group D	38 + 9	39 + 8	18 + 4	21 + 6
30 days	Group A	40 + 5	30 + 5	15 + 2	25 + 5
	Group B	45 + 5	34 + 3	17 + 4	25 + 3
	Group C	48 + 7	41 + 4*	14 + 3	26 + 2
	Group D	42 + 4	29 + 5	15 + 4	23 + 4

Proliferative responses (cpm)

	PHA	Con A	PWM	
Group A	84897 + 20026	88498 + 27971	24099 + 7174	15 days
Group B	76474 + 12844	78918 + 20394	27873 + 5007	
Group C	126320 + 19969**	123919 + 37750**	43341 + 11850**	
Group D	80981 + 18055	92704 + 24684	23976 + 9861	
Group A	58184 + 12157	62188 + 12583	32291 + 9058	30 days
Group B	56943 + 11621	63481 + 14712	35894 + 8612	
Group C	63517 + 10687	84570 + 13750*	41595 + 4684*	
Group D	56660 + 13978	61869 + 12772	31754 + 7352	

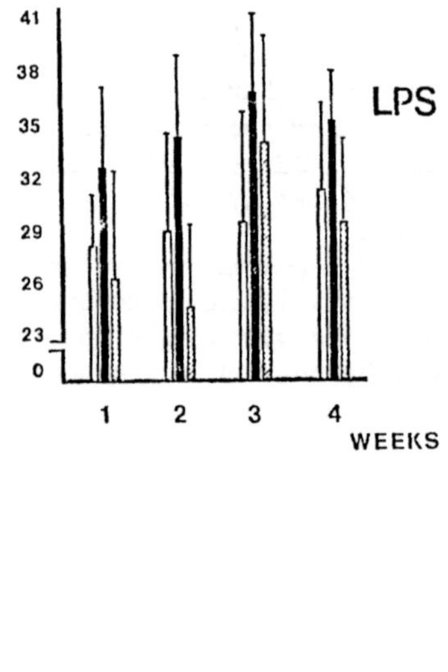

LA VISCOSITE COMME EVALUATION DE LA TEXTURE DU YAOURT BRASSE

R. CORNET et M. COLOMBINI
Laboratoire Control Qualité, Yomo, Milano, Italie

INTRODUCTION

Une recherche a été effectuée pour évaluer la variation de la viscosité du yaourt brassé industriel produit avec technologie inchangée au cours d'une année afin de fournir un standard qualitatif.

MATERIAUX ET METHODES

216 échantillons de yaourt, obtenu du lait entier frais, dérivants de 9 productions mensuelles au cours d'un an ont été soumis à un examen analytique:

- au moment de la fabrication détermination du résidu sec total et du résidu sec maigre du lait utilisé.

- 24 heures après la production du yaourt détermination de :
  a) Viscosité a 4°C avec viscosimètre rotatoire Haake, système RV 12 à cylindre coaxial programmé à 64 tours par minute. La lecture de référence était effectuée à moitié du trace viscosimètrique et exprimée en cP (centipoise).
  b) Acidité en pourcentage d'acide lactique
  c) Dénombrement des bactéries lactiques spécifiques selon FIL-IDF 117:1983
- après 30 jours de stockage à 4°C :
  a) Contrôle organoleptique du produit
  b) Acidité en pourcentage d'acide lactique
  c) Dénombrement des bactéries lactiques spécifiques

DISCUSSION

- Il y a correspondance entre la viscosité du yaourt et l'extrait du lait.
- aux valeurs les plus basses de viscosité, toutefois, le yaourt ne présente pas de sinerese même après 30 jours de conservation à 4°C.

- à la production, l'acidité du yaourt était inclue entre 1.10-1.25% d'acide lactique et la flore spécifique entre 10^8 - 10^9/g; à 30 jours après la fabrication l'acidité était de 1.30 - 1.35% et le nombre des bactéries lactiques de 10^7 - 10^8/g.

CONCLUSION

Les valeurs de viscosité trouvées, même avec delta significatif, représentent des paramètres valides conjointement à la flore lactique spécifique pour l'évaluation qualitative du yaourt brassé de lait entier.

En particulier, la technologie utilisée compense de façon satisfaisante les éventuelles variations imputables aux plus basses teneurs en solides du lait et à une viscosité du yaourt qui est, par consequent, moins élevée.

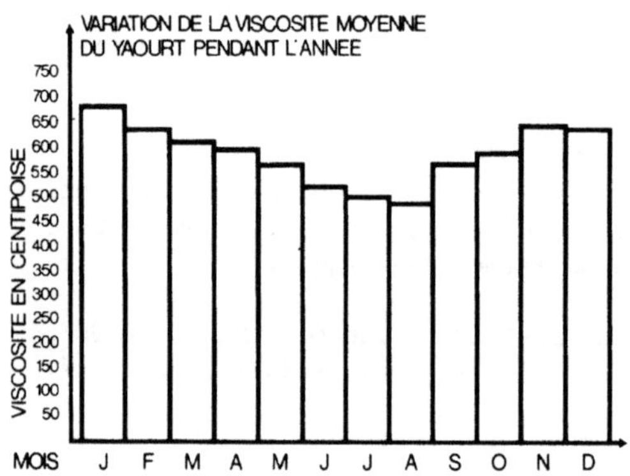

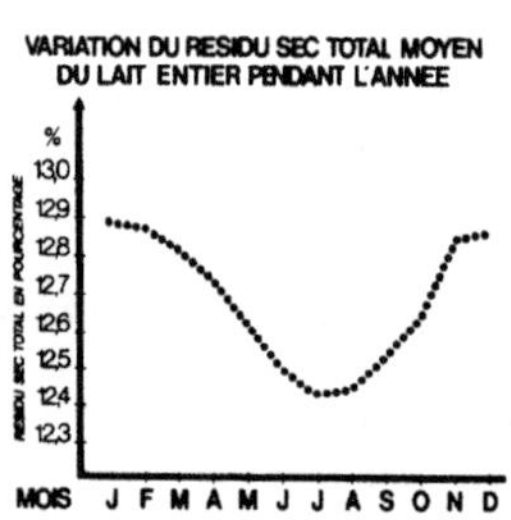

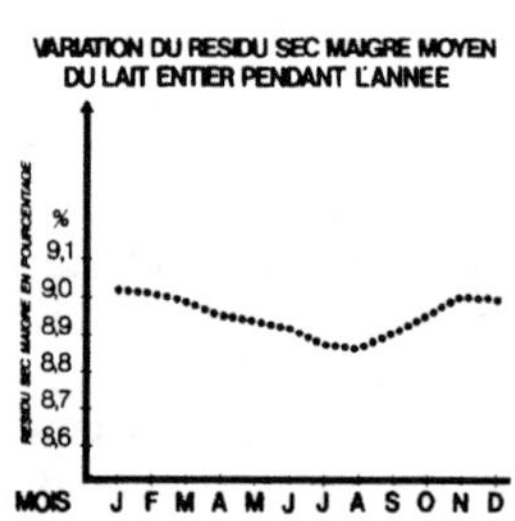

SEPARATION OF PEPTIDASES USING FAST PROTEIN LIQUID CHROMATOGRAPHY (FPLC)

YLVA ARDÖ
Swedish Dairies' Association (SMR),
Central Laboratory, Box 205, 201 22 Malmoe, Sweden

ABSTRACT

Lactic acid bacteria produce a wide range of peptidases with various specificity in hydrolysing peptides. To study the peptidolytic properties of cheese bacteria a simple method for enzyme separation was developed. Using anion exchange chromatography (Mono Q, FPLC, Pharmacia) the peptidases in crude extracts were separated in 30 min. Their enzymatic activities were recovered and specific activities against various substrates were studied.

MATERIAL AND METHODS

Starter streptococci were grown in the same way as for cheese making. The culture was then neutralized, the bacteria were washed and an ultrasonic extract was freeze dried. A solution of this is called the crude extract.

FPLC is a fast liquid chromatography system especially designed for separation of proteins, peptides and polynucleotides. The system has an interesting series of columns containing different gel materials. An anion exchange column (Mono Q, 1 ml) was tested. Gel filtration in Sephacryl S300 was performed using the same equipment. The eluation methods used are presented in the figures.

Peptidolytic activities were measured using a modification of the method of Lewis and Harris (1967). The o-dianisidin hydro chloride was replaced with 2,2'-azinodi-(3 ethylbenzthiazoline (ABTS) and the reactions took place in test tubes. The activities were tested against phenylalanylleucine and glycyl-leucine for dipeptidases, leucine p-nitroanilide (LNA) and proline p-nitroanilide (PNA) for aminopeptidases and carbobenzoxy-alanyl-phenylalanine and carbobenzoxy-isoleucyl-proline for carboxypeptidases. Hydrophobic substrates were used in the purpose of studying debittering properties of the peptidases.

RESULTS AND DISCUSSION

The peptidases were successfully separated with anion exchange chromatography on FPLC. The activities were recovered and could be characterized. A separation of the crude extract gave valuable information for classifying of the cheese bacteria after their peptidolytic properties.

ACKNOWLEDGEMENTS

The author is indebted to Mrs Lena Jönsson and Mr Hiroshi Ogura for excellent technical assistance.

REFERENCES

Lewis, W.H.P. & Harris, H. 1967. Human red cell peptidases. Nature 215, 351-355.

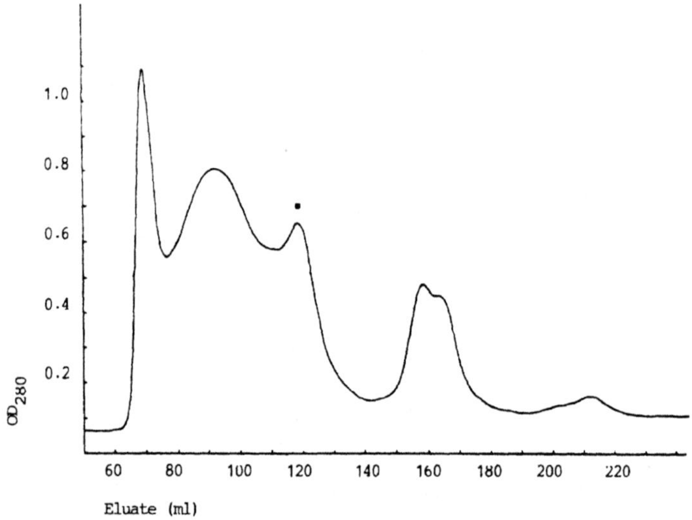

Figure 1. Gel filtration of crude extract on Sephacryl S300. Buffer: 0.1 M sodium phosphate, pH 7.0. Sample: 40 mg in 1 ml buffer. Flow rate: 30 ml/hour. Gel volyme: ca 130 ml.

- Peak with the most peptidolytic content - Dipeptidase, aminopeptidase and carboxypeptidase.

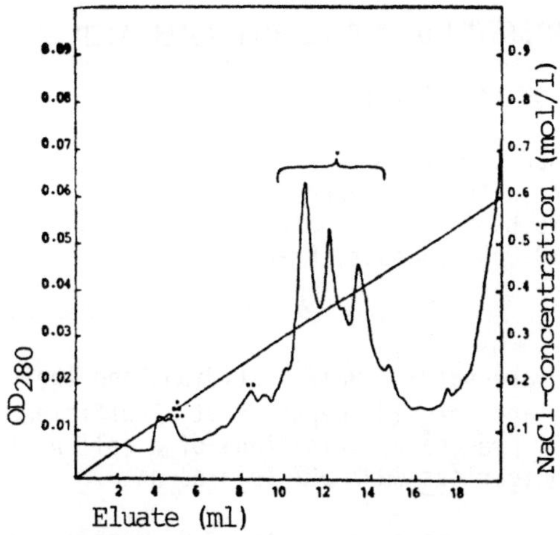

Buffer: 20 mM TRIS pH 7.6
Flow rate: 1 ml/min
Gradient: 0 - 0.6 M NaCl
Gel volyme: 1 ml
Sample size: 2.5 ml

Activities:
· Dipeptidase
·· Aminopeptidase
··· Carboxypeptidase

Figure 2. FPLC-separation on Mono Q of materials from the peak with the most peptidolytic content in figure 1.

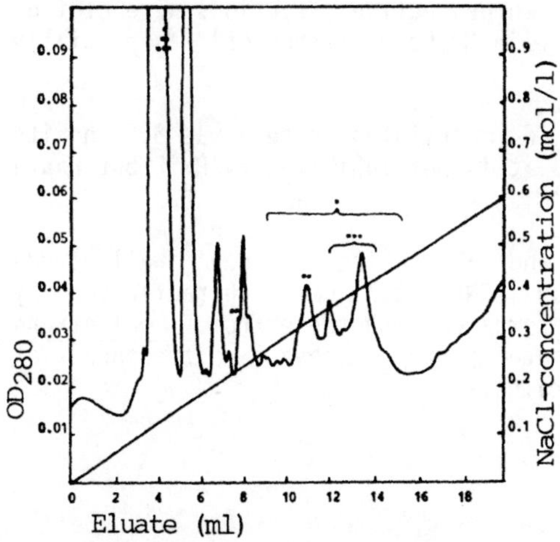

Buffer: 20 mM TRIS pH 7.6
Flow rate: 1 ml/min
Gradient: 0 - 0.6 M NaCl
Sample: 2 mg in 0.5 ml buffer

Activities:
· Dipeptidase
·· Aminopeptidase
··· Carboxypeptidase

Figure 3. FPLC-separation on Mono Q of crude extract.

PARTIAL CHARACTERISATION OF A CELL WALL ASSOCIATED PROTEINASE FROM LACTOBACILLUS BULGARICUS

N. EZZAT[1], C. ZEVACO[2], M. EL SODA[1], J.C. GRIPON[2].
1. Faculty of Agriculture, Alexandria, Egypt.
2. Laboratoire de Biochimie et Technologie Laitières, INRA, 78350 Jouy-enJosas, France

Lactic acid bacteria are nutritionaly fastidious and their proteinases hydrolyse casein and provide peptides at a sufficient level for maximal growth. A pre-characterisation of a cell wall associated proteinase from L. bulgaricus CNRZ 397 is presented.

Proteolytic activity was measured with ^{14}C methylated casein. Cells were grown on MRS broth for 11h at 42°C and pH was maintained constant at a value of 6.0. Cell wall associated proteinases were released from centrifuged cells by incubation at 25°C in Tris-HCl buffer 0.05 M pH 7.8. Four successive extractions (1 hour each) released respectively 7560, 5400, 5400 and 1120 proteolytic units. Total release of intracellullar lactate-deshydrogenase activity in the extracts was lower than 4 % and indicated little intracellular leakage.

The cell wall crude extract was partially purified by DEAE-Sephacel chromatography. Three active fractions were eluted. The major one (eluted with 0.135 M phosphate buffer) was partially characterized.

Its optimal pH on ^{14}C methylated casein was 5.5 and its optimal temperature was 35°C. It is not inhibited by PMSF but has a 32 % inhibition with EDTA 10^{-3}M.

Hydrolysates of αs1 and β-caseins by cell wall crude extract from L. bulgaricus CNRZ 397 were characterized by isoelectric focusing and were compared with those from L. helveticus CNRZ 303. β-casein hydrolysate patterns reveal differences of specificity between the two strains.

Effect of Rennet Level on Bitterness Development in Quarg

T.S. Sohal and P. Jelen
Department of Food Science
University of Alberta
Edmonton, Alberta, T6G 2P5
Canada

Development of bitterness in quarg could be a problem limiting the product shelf life. It is known that rennet (1) and/or starter bacteria (2) can produce bitterness in hard cheeses, while the mechanism of bitterness development in quarg is less understood. Studies were undertaken to elucidate and rectify the problem of bitterness development in commercial quarg manufactured in Alberta by a traditional technology including acid coagulation of milk with small amounts of rennet and whey separation by a double bottom perforated vat. Five batches of quarg in duplicate were produced in laboratory using 0, 388, 775, 1550 and 3876 units of rennet per 1000 kg of milk and 1% starter culture (Flora Danica). Quarg samples were stored at 7°C for 4 weeks and analysed at weekly intervals to follow the effect of storage on predominant microorganisms, pH and lactic acid content. In addition, the samples were scored at weekly intervals by a panel of six trained judges on a 1 to 6 bitterness scale. The analysis of variance performed on the sensory data showed highly significant difference among the rennet levels affecting bitterness development. A comparison of means by Tukey's test indicated that results with 0 and 388 rennet units per 1000 kg of milk were significantly different (5% level) from 775, 1550 and 3876 units per 1000 kg of milk. The results of storage study indicated low level of psychrotrophs, yeasts and molds, suggesting that the bitterness development in quarg was not the outcome of microbial contaminants. A separate quarg-making experiment with 388 rennet units and four different starters (Flora Danica with and without *S. diacetilactis*, and *S. cremoris* single strains #134 and #584) indicated no major effect of the starter cultures. Though bitterness in quarg was at a minimum when no rennet was used, the yield was poor compared to the quarg produced with 388 or more rennet units.

References

1. Stadhouders, J. *et al*. Neth. Milk Dairy J. *37*:157.
2. Lawrence, R.C. *et al*. 1972. N.Z.J. Dairy Sci. Technol. *7*:32.

ACID PRODUCTION OF MIXED CULTURE OF LACTOBACILLUS BULGARICUS AND STREPTOCCUS THERMOPHILUS IN SKIM MILK RETENTATES.

S.EL-SHIBINY, M.B. MAHFOUZ, H.F. EL-DEIN and M.H. ABD EL-SALAM
National Research Centre, Cairo, Egypt

The effect of quantity of starter (1-10%), addition of NaCl (0-5%) on acid development by mixed starter of Lactobacillus bulgaricus and Streptoccus thermophilus in skim milk retentates containing 4, 8, 12 and 17.3% protein was studied. Increasing the quantity of starter and protein content stimulated acid development by starter bacteria. Addition of 2-3% NaCl inhibited acid development in retentates containing 12 and 17.5% protein.

INTRODUCTION

Several cheese varieties and fermented dairy products have been made by ultrafiltration techniques on pilot and industrial scale (Glover et al. 1978: Maubois, 1980). However, few studies have been undertaken to study the behaviour of starter bacteria in UF milk retentates (Hickey et al.1983). Preliminary study on the production of acid coagulated skim milk cheese (Abd El-Salam et al.1984) by UF techniques showed that the growth and acid production by yoghurt starter in retentates was sensitive to NaCl concentration.
The present paper deals with the behaviour of a mixed culture of Lactobacillus bulgaricus and Streptoccus thermophilus in skim milk retentates as affected by protein content, quantity of inoculum and sodium chloride added.

MATERIALS AND METHODS

Reconstituted skim milk (10% TS) in deionized water

Ultrafiltration 1-Equipment used: DDS Lab-20 (2 m^2)
2-Conditions: at 45°C, pressure inlet 3.6 bar outlet 0.6 bar, concentration factor .4.

Retentate: heated to 70°C, protein content adjusted to 4,8,12 and 17.3% using permeate.

Culture used mixture of L.bulgaricus and S.thermophilus(1:1)

Treatments 1- quantity of inoculum 1,2,3,5 and 10% of the mixed culture

2- NaCl added, 0, 0.5, 1.0, 2.0 and 3.0%

Inoculated retentates from different treatments were incubated at 37°C and titratable acidity was followed every 2 h up to 8 h. Results are expressed as increment increase in acidity.

RESULTS AND DISCUSSION

Fig 1 to 4 show the developed acidity by mixed culture of L. bulgaricus and S. thermophilus in skim milk retentates containing 17.3, 12, 8 and 4% protein respectively as affected by quantity of inoculum and NaCl added.

Effect of added NaCl on developed acidity was affected by the protein content of retentate. 0.5% NaCl stimulated acid development by mixed culture in 17.3% protein retentate, while the addition of 1% NaCl to 12 and 17.3% protein retentate slightly decreased the rate of acid development. At 2-3% NaCl level in retentates with high protein content (12 and 17.3%) acid development was almost inhibited.

In retentates with low protein content (4 and 8%), addition of 1% NaCl stimulated acid development by starter bacteria. Addition of 2-3% NaCl to these retentates slowed down acid development by starter bacteria.

Effect of protein content In retentates containing 0.5% NaCl a linear and positive relation between developed acidity and protein content was apparent after 4 h which was probably due to

1- Increased buffering capacity of the medium as the protein content increased.
2- Increased retention of some trace elements sequestered to proteins.
3- Increased whey protein stimulated the growth and activity of starter bacteria (Marshall et al. 1982).

Effect of the quantity of starter. Acid development was enhanced by the increase in the quantity of starter used. This could be explained that increasing the quantity of starter shortens the lag phase of starter growth (Davis,1965).

The foregoing results indicate that the protein content of the retentate, quantity of starter used and addition of NaCl were interrelated in their effect on acid development by L. bulgaricus and S. thermophilus.

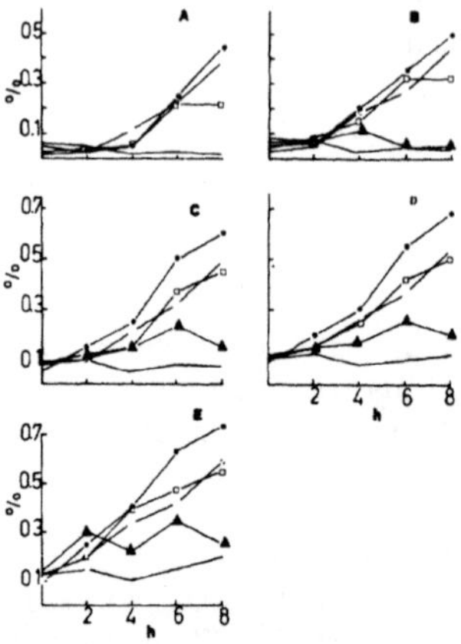

Fig 1: Developed acidity by mixed culture of L.bulgaricus and S. thermophilus in skim milk retentate containing 17.3% protein and inoculated with (A) 1% (B) 2%, (C) 3%, (D) 5% (E) 10% starter as affected by added NaCl: □, without; ○, 0.5%; ●, 1%; ▲, 2%; ·, 3%.

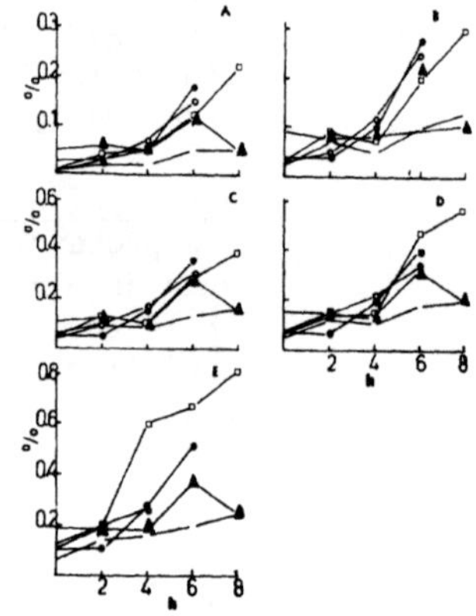

Fig 2: Developed acidity by mixed culture of L.bulgaricus and S. thermophilus in skim milk retentate containing 12% protein inoculated with (A) 1%, (B) 2%, (C) 3%, (D)5%, (E) 10% starter as affected by added NaCl: □, without; ○, 0.5%; ●, 1%; ▲, 2%; ·, 3%.

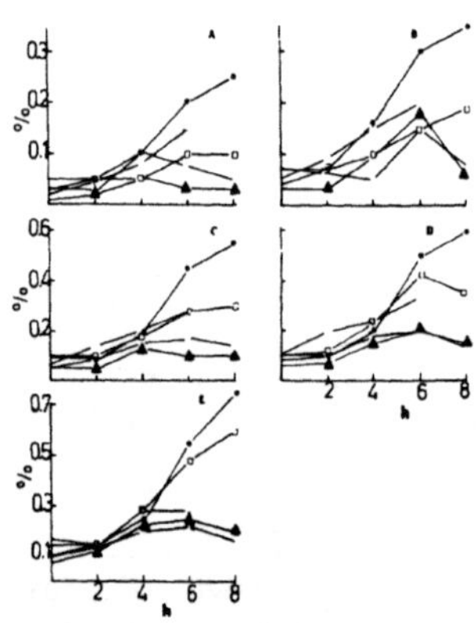

Fig 3: Developed acidity by mixed culture of L. bulgaricus and S. thermophilus in skim milk retentate containing 8% protein and inoculated with (A) 1%, (B) 2%, (C) 3%, (D) 5%, (E) 10% starter as affected by added NaCl: □, without; ○, 0.5%; ●, 15%; ▲, 2%; ·, 3%.

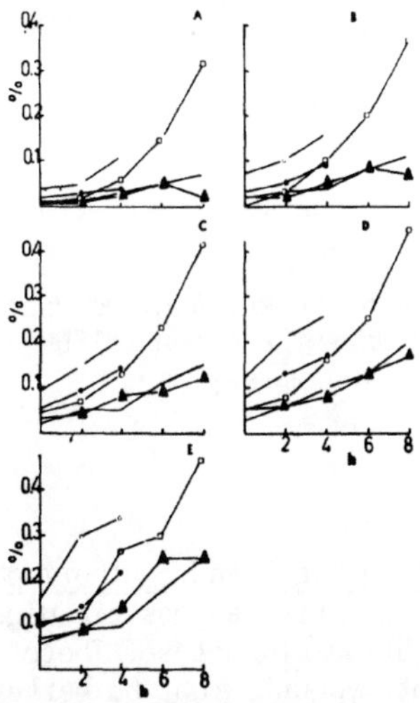

Fig 4: Developed acidity by mixed culture of L. bulgaricus and S. thermophilus in skim milk retentate containing 4% protein inoculated with (A) 1%, (B) 2%, (C) 3%, (D) 5%, (E) 10% starter as affected by added NaCl:□, without;○, 0.5%;●;1%; ▲,2%; · ;3%.

REFERENCES

Abd El-Salam, M.H., El-Shibiny, S., and El-Alamy, H.A., (1984). Egyptian Journal of Dairy Science, 12 : 11.

Davis J.G. (1965). "Cheese Vol 1 Basic Technology" Churchill Ltd. London

Glover I.A., Skudder P.J., Stothart C.H. and Evans I.W. (1978) Journal of Dairy Research, 43, 291

Hickey M.W., Roginski M., and Bromer M.C., (1983) Australian Journal of Dairy Technology, 18, 138

Marshall V.M., Cole W.M. and Vega N.J. (1982). Journal of Dairy Research, 49, 665.

Maubois J.L., (1980)"Recent developments of membrane ultrafiltration in dairy" in "Ultrafiltration membranes and application"
Id. A.R. Cooper, Plenim Pub. Co.

NUTRITIONAL ASPECTS OF YOGURT. I. MICROBIAL LACTASE ACTIVITY AND DIGESTION OF LACTOSE

G. Schaafsma, P. Derekx, P.R. Dekker & H. de Waard
Netherlands Institute for Dairy Research (NIZO)
P.O. Box 20, 6710 BA EDE, The Netherlands

Milk fermented by a culture of L. bulgaricus and S. thermophilus (yogurt) shows microbial lactase activity which survives gastric digestion and might contribute to intestinal digestion of lactose. Therefore and because of its lower lactose content, yogurt might be better tolerated by lactase-deficient subjects than milk. The significance of lactase in yogurt was investigated in rats which are known to be lactase-deficient. Five groups of rats were fed, for six weeks diets based on milk (M, n = 12), lactase-treated milk (L, n = 12), yogurt (Y, n = 12), pasteurized yogurt (P, n = 12) or a commercial rat diet (C, n = 6). Rats fed Y or P did not show significant differences in feed efficiency, blood galactose response and relative caecum weight, indicating that microbial lactase in Y did not contribute significantly to the digestion of lactose in vivo. Rats fed milk showed a weak blood galactose response, a reduced food consumption and weight gain and an increased relative caecum weight as compared to the other groups. These results are in agreement with the experience in lactase-deficient subjects that yogurt and lactase-treated milk are better tolerated than milk.

NUTRITIONAL ASPECTS OF YOGURT. II. INTESTINAL ABSORPTION OF MINERALS

G. Schaafsma, P.R. Dekker & H. de Waard
Netherlands Institute for Dairy Research (NIZO)
P.O. Box 20, 6710 BA EDE, The Netherlands

An increase in the per capita consumption of cultured dairy products, particularly yogurt, has been noticed in the recently passed years in Europe. Like milk yogurt is an excellent source of essential nutrients (proteins, vitamins and minerals). It is not clear from the available literature whether the bioavailability of minerals from yogurt is comparable to that from milk. Lactose is known to improve the intestinal absorption of calcium, magnesium and probably zinc. Thus a decrease of the lactose content following fermentation might result in a decrease of the bioavailability of these nutrients. On the other hand it has been speculated that the organic acids, produced during fermentation, may have a favourable effect on mineral absorption. In the present study the effect of fermentation and lactose hydrolysis of milk on the intestinal absorption of calcium, magnesium, zinc and phosphorus has been studied in rats. It appeared that both treatments of milk led to changes in the bioavailability of these minerals.

Rats fed yogurt, pasteurized yogurt or lactose-hydrolysed (70 %) milk showed a significantly lower intestinal absorption of calcium, magnesium and zinc than rats fed milk. This reduction in intestinal mineral absorption was correlated with the decrease of the lactose content of the diets following fermentation or lactase treatment. The intestinal absorption of phosphorus from both types of yogurt was significantly higher than that from milk or lactose-hydrolysed milk.

STABILITY AND ACTIVITY OF BD-TYPE STARTER IN CONTINUOUS CULTURE

Z. Libudzisz, A. Piatkiewicz, H. Oberman
Institute of Fermentation Technology and Microbiology, Technical University, 90-924 Łódź, B. Stefanowskiego 4/10, Poland

The continuous culture conditions have been elaborated, giving possibility of mixed starter biomass production, especially for the preparation of concentrated mixed bacterial starters.

BIOLOGIC ENRICHMENT OF MILK PRODUCTS WITH VITAMINS

J.Černá, H.Hrabová[x], B.Dryák[xx], M.Dědek
Research Institute of Food Industry, Prague-5, Smíchov
Na Bělidle 21, 150 38 - ČSSR
x Dairy Research Institute, Prague-9, Kyje,Českobrodská 1174, 198 00-ČSSR
xx Concern PMV-Nový Bydžov, 504 01-ČSSR

A special strain of Propionibacterium shermanii (P.sh.) was shortly activated (2 x 6 h at 30 °C) in a semisynthetic medium and used either alone or in symbiosis with a lactose fermenting yeast,e.g. Kluyveromyces fragilis (K.f.). The contents of vitamin B_{12}, folacin, thiamine, niacin and vitamin B_6 were thus significantly enhanced in several milk products. The two microorganisms were first cultivated separately in sterile milk to prepare a "milk inoculum". In the case of P.sh., 5 % of the activated inoculum and in that of K. f. only 1 % of a yeast milk culture (24 h at 30 °C) were used. Depending on the milk product required, either 1 % of this P.sh. inoculum alone, or together with 1.0 or 0.5 % of K. f. milk inoculum was used simultaneously with the cream starter.

The concents of vitamins in different enriched products, i.e.: fermented milk beverage, thermised curd, thermised cream cheese and two typ es of powdered milks, are presented in graphs 1 - 5 in comparison with those in non-enriched products. The highest enrichment with vitamins was achieved in powdered milks, where the negative vitamin-consuming effect of the cream-starter bacteria did not occur. The rather unpleasant taste and flavour of propionic acid in these products was completly eliminated by the high temperature used during spray drying. On the other hand, the unpleasent taste and flavour of yeasts in the case of curd and cream cheese disappeared at temperatures of about 75 °C and high pressure during the process of thermisation. The taste of yeasts in milk beverage was acceptable and remindsthat of kefir.

In all the products, the highest increase was observed in the concents of vitamin B_{12} and folacin, which was due to the biosynthesis of P.sh. These vitamins are produced in natural--occuring protein-bound forms and are consequently readily absorbed. These forms exhibit a much greater vitamin activity than the pure vitamin preparations.

The production of an enriched milk beverage called Elvit has beenalready started in ČSSR. The production of an enriched sort of a thermised cream-cheese is under preparation.

YOGURT MADE FROM MILK WITH ADDED LACTULOSE

A.Olano, M.Ramos, J.A.Suarez, S.L.Lopez-Covarrubias

Instituto de Fermentaciones Industriales (CSIC)
Juan de la Cierva, 3. 28006 Madrid. (Spain)

The object of this work was to study the effect of the addition of lactulose during the manufacture of yogurt or yogurt in which part of the lactose was hydrolysed by using Maxilact.

Lactulose was reduced by 30% when milk containing 3,6% added lactulose was incubated with Maxilact to achieve 50% lactose hydrolysis. The presence of lactulose did not affect the fermentation rate which was followed by measuring the pH and lactic acid formation at intervals.

Lactulose concentration decreased during fermentation even when single starter organisms were used (L. bulgaricus being cultured in milk containing 50 ppm Na formate). When Maxilact was added together with the starter, lactulose was reduced by 35-40% after fermentation. In absence of Maxilact, lactulose decreased by 11-25% after fermentation.

The proteolysis of yogurt (measured as trichloroacetic acid-soluble products) was not affected by the presence of lactulose. The free aminoacids (sulphosalycilic acid -soluble products) were slightly higher in yogurts prepared from hydrolysed milk. Polyacrylamide gel electrophoresis of Caseins showed no degradation after fermentation in all samples.

Lactulose is hydrolyzed by Maxilact resulting in a mixture of galactose and fructose with a sweetness level higher than that of hydrolysed lactose due to the presence of fructose. The incorporation of lactulose into yogurt can be of considerable interest, particulary in dietetics and in the nourishment of people suffering intestinal infections.

YOGHURT FROM UHT MILK

Ajay Goyal and D.N. Prasad
Ultrastructure Lab., Dairy Microbiology Division,
National Dairy Research Institute, Karnal - 132001, India

UHT - milk is gaining popularity in India due to its long shelf life and hence there is a need to investigate the suitability of UHT - milk for conversion to fermented products. In the present study, efforts have been made to convert UHT - milk to Yoghurt with low acidity, increased flavour with a long shelf-life. Yoghurt was made using Lactobacillus bulgaricus (RTS) and Streptococcus thermophilus (HST). Observations were made on the growth pattern, acidity and production of Carbonyl Compounds on storage at 4°-6°C upto one week. The yoghurt was compared with that prepared from whole Cow milk. Product made from UHT - milk showed firm body and texture with prominent aroma and higher volatile acidity. During storage, volatile and titratable acidities showed a gradual increase while decrease in acetaldehyde content was observed. However, the Acetaldehyde was comparable in both Cow and UHT milk yoghurts. Cell counts of Lactobacillus bulgaricus were higher in Cow milk yoghurt than in UHT milk yoghurt.

SCANNING ELECTRON MICROSCOPY OF INDIAN DAIRY PRODUCT - DAHI

D.N. Prasad and N. Prasad
Dairy Microbiology Division, National Dairy Research Institute, Karnal - 132001, India

Dahi, an important fermented milk product with good taste, flavour, high nutritive and therapeutic value is consumed throughout India. Most of the earlier studies were confined to microbiological and chemical aspects but no information is available on the microstructure of Dahi coagulum. The present study was undertaken to examine the interaction of milk casein and lactic microorganisms in Dahi prepared from Buffalo milk with the help of Scanning Electron Microscope. Dahi sample was prepared from Buffalo milk steamed for 20 min. and cooled to 25°C followed by addition of 1% LF-40 starter. It was thoroughly mixed and incubated at 25°C for 12 hours. Small cubes of 2mm x 2mm x 2 mm, were cut from fresh curd, fixed in 2.5% glutaraldehyde at 4°C for 1-2 hours and doubly fixed in 1% O_sO_4 for 30 min. at low temperature. These were then sputtercoated with gold around 200 A° in thickness and examined under Hitachi S-405A SEM at 15 KV. The microstructure of curd coagulum revealed a fascinating view of compact matrix with small cavities. It presented a uniform aggregation of casein micelles. Presence of streptococci in short chains or singly were observed evenly throughout the matrix. Normally these starter microorganisms were localised in or around deep pockets. A number of well preserved spherical fat globules were also seen embedded in the casein background. The study of microstructure therefore could be reliably employed as an index of quality in terms of the body and texture of the product.

L'étude des effets antibactériens de l'aflatoxine M_1 caracterise le comportement de souches Lactobacillus, ou résistents. L'alteration de la sensibilité des souches Lactobacillus est sous la dependance de l'effet concentration de l'aflatoxine M_1 Sensibilité de souches de lactobacillus sensibles et resistentes a l'aflatoxine M_1.

Stojanka Mitić, M.Rakić, Anka Kasalica

Institute de véterinaire et laiterie - Belgrade Autoput 3, Yugoslavie

Trente souches suivantes ont été utilisées: Lactobacillus helveticus, L.bulgaricus, L.acidophilus at L.jugurti, dont les caracteristiques sont identifiées par API sistem.
L'activité antbacterienne de l'aflatoxine M_1 est mise en évidence par la méthode de dulition en milieu liquide. Les souches'étaient dans le lait écremé exempt d'antibiotiques. On ajoute 1 ml d'aflatoxine M_1 dans chlorophorm solutions (o,o5 - o,2 ug/ml). Les tubes de lait ensemencés étaient incubées a 37°C pendant 24, 48 et 72h. Les résultets sont quantifies par évaluation d'inhibition de croissance et de l'activité acidifiante de bactéries lactiques.
L'ensemble des résultats montrent que tous les souches sont inhibées par de toxique de o,2 ug/ml, 7o% par de o,1 ug/ml, 3o% par de o,15 ug/ml et 2o% par de o,o5 ug/ml. Les plus sensibles sont des souches de L.bulgaricus. Ce relantissement de croissance s'accompagne de l'apparation de cellules anormalement allongées et de cellules atrophiées.

Biological transformation of whey for obtaining refreshment drinks

Mr Dragica Miočinović, Dr Stojanka Mitić, Anka Kasalica,dipl.ing
Veterinary and Dairy Research Institute - Department of Dairy Research - Novi Beograd, Auto put 3 - Yugoslavia

The whey is by product of cheese-making and it's been treated like an unusefull product in milk production. In order to better usses of nutritive value in feeding adults and children the whey should be treated by a temperature. Biochemical analysis has to be done too. For experiments, has been used selective cultures of lactobacillus, which have been shown great antimicrobic action on pathogenic and potentially pathogenic Enterobacteriaceae. Concentrated fruits has been added in whey drink for increasing organoleptic qualities (peaches,cherries etc.).

On occasion of production has been obtained fermented whey drink riched in protein and almost all aminoacids which has given its high biologycal value. Whey drink is also riched in important mineral components for human organism, and also in other proteins (albumine and globuline).

The whey drink has had good organoleptic effects which confirms that the technological operations of obtaining fermented whey whith using L.helveticus whith potential antimicrobicall effect has been overmastered.

On that way the nutritive value of whey is connected whith inhibitory factors of starters against pathogenic and potentionally pathogenic and gram-positive bacteries.

On occasion of production has been obtained a good quality whey by the selection of the Lactobacillus helveticus, which has antimicrobical effects. The complete chemical and microbiological analysis of the drink has been done. It's been found almost all esential aminoacids.

FORMATION OF FLAVOUR AND AROMA IN TURKISH YOGHURTS

Kaptan N. and Gürsel A.
Ankara University, Faculty of Agriculture
Dairy Technology, Ankara, Turkey

To search the effects of technological developments in manufacturing of yoghurt, two strains each of Streptococcus thermophilus and Lactobacillus bulgaricus, were chosen according to their acid and acetaldyde production and they were combined in pairs.

Result
It was found that the usage of S.thermophilus 1_{11}+L. bulgaricus G. will meet the consumers demand throughout Turkey.

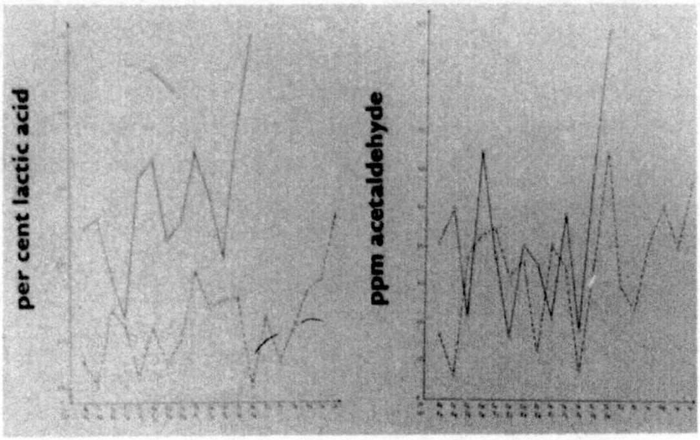

Titratable acidity   Acetaldehyde production

After twenty four hours of incubation at 42C

(1) Strains of S.thermophilus Δ----Δ by S.thermophilus

(2) Strains of L.bulgaricus •——• by L.bulgaricus

Acid and acetaldehyde production were determined in the samples obtained by using these cultures.

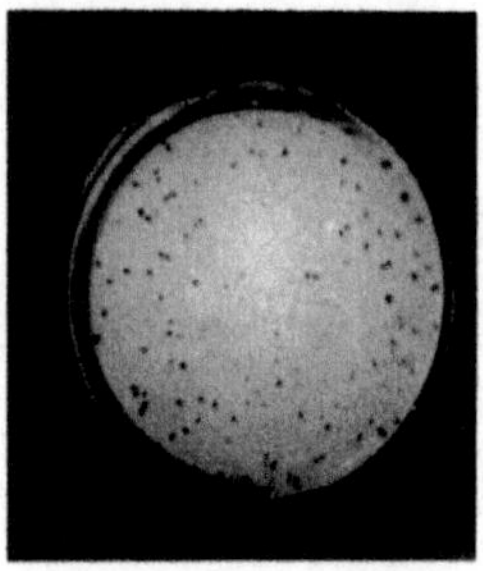

Lactobacillus bulgaricus G.

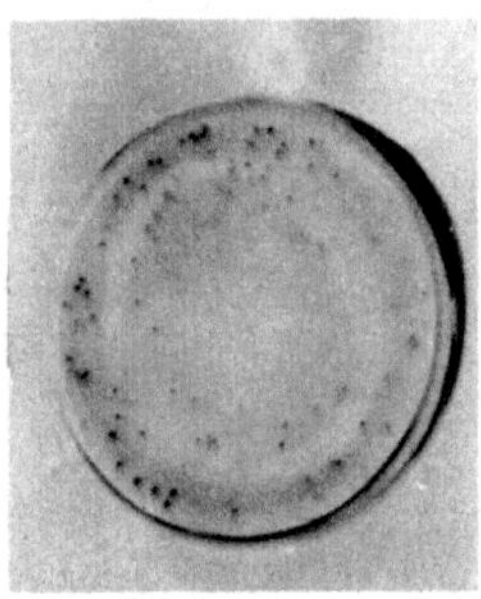

Streptococcus thermophilus

Acidity, pH and acetaldehyde in the yoghurt samples

1 S.thermophilus E_5 + L.bulgaricus C_8
2 S.thermophilus E_5 + L.bulgaricus G_6
3 S.thermophilus 1_{11} + L. bulgaricus C_8
4 S.thermophilus 1_{11} + L. bulgaricus G_6

—•— Titratable acidity
•—• Acetaldehyde
..... pH

ORGANIC ACIDS AND VOLATILE AROMA COMPOUNDS IN GOAT'S MILK FERMENTED WITH SINGLE STRAINS OF MESOPHILIC LACTIC ACID BACTERIA

Gunnar Rysstad and Roger K. Abrahamsen
Department of Dairy and Food Industries,
Agricultural University of Norway,
N-1432 Aas-NLH, Norway

INTRODUCTION

PRODUCTION OF VOLATILE AROMA COMPOUNDS AND CERTAIN ORGANIC ACIDS IS IMPORTANT FOR THE QUALITY OF FERMENTED MILK PRODUCTS. DIFFERENT PRODUCTS ARE CHARACTERIZED BY THE CONCENTRATION AND RELATION BETWEEN SUCH METABOLITES CAUSED BY BACTERIAL FERMENTATION.

EARLIER INVESTIGATIONS (RYSSTAD AND ABRAHAMSEN, 1983) SHOWED THAT THE PRODUCTION OF DIACETYL, α-ACETOLACTIC ACID AND ACETOIN WAS MUCH LOWER IN GOAT'S MILK THAN IN COW'S MILK FERMENTED WITH MESOPHILIC, MIXED STRAIN STARTERS. THE FORMATION OF ETHANOL, HOWEVER, WAS DISTINCTLY HIGHER IN GOAT'S MILK.

THE PRESENT STUDY WAS UNDERTAKEN TO INVESTIGATE PRODUCTION OF FLAVOUR COMPOUNDS IN COW'S AND GOAT'S MILK FERMENTED WITH SINGLE STRAIN STARTERS.

CONCLUSION

RESULTS FROM THE PRESENT INVESTIGATION LEADS TO THE FOLLOWING CONCLUSIONS

- THE CONTENT OF CITRATE IS LOWER IN GOAT'S MILK THAN IN COW'S MILK
- GOAT'S MILK FROM THE LATTER PART OF LACTATION CONTAIN SIGNIFICANTLY LESS CITRATE THAN MILK FROM EARLY LACTATION
- THE CONTENT OF ORGANIC ACIDS AND VOLATILE AROMA COMPOUNDS VARIED IN THE THREE MILKS INCUBATED WITH THE SAME BACTERIAL STARTER
- LC. CREMORIS IS RESPONSIBLE FOR THE HIGH PRODUCTION OF ETHANOL OBSERVED IN GOAT'S MILK, PARTICULARLY IN GII

ALL RESULTS INDICATE THAT GOAT'S MILK FROM DIFFERENT STAGES OF LACTATION WILL GIVE FERMENTED PRODUCTS WITH VARIABLE CONTENT OF AROMA COMPOUNDS

MATERIALS AND METHODS

MILKS
SAMPLES OF GOAT'S MILK FROM EARLY (GI) AND LATE (GII) LACTATION WERE SKIMMED AND DRIED. COW'S MILK WAS USED AS A REFERENCE MEDIUM. ALL MILKS WERE RECONSTITUTED (10% W/V) AND HEAT-TREATED (115°C FOR 5 MIN) BEFORE USE IN THE EXPERIMENTS.

CHANGES IN THE MILK CITRATE CONCENTRATION DURING THE LACTATION PERIOD OF GOAT (UNIVERSITY HERD, 70 GOATS) IS PRESENTED IN FIGURE 1.

FIG. 1. CHANGES IN MILK CITRATE CONCENTRATION DURING THE LACTATION PERIOD OF GOAT (UNIVERSITY HERD, 70 GOATS)

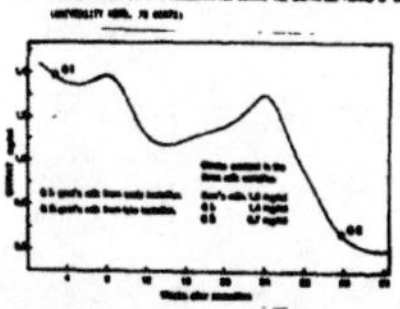

CHEMICAL ANALYSIS
ORGANIC ACIDS WERE ANALYSED BY HPLC AFTER EXTRACTION WITH ACETONITRILE.
ALL SAMPLES WERE ACIDIFIED WITH H_2SO_4 (PH=2.5) TO OBTAIN COMPLETE EXTRACTION OF CITRIC ACID.
VOLATILE AROMA COMPOUNDS WERE ANALYSED BY GAS CHROMATOGRAPHY AFTER DISTILLATION IN OXYGEN AND NITROGEN ATMOSPHERE.

CULTURES
THREE DIFFERENT MESOPHILIC SINGLE STRAIN BACTERIA WERE INVESTIGATED.
LC. CREMORIS PD-1 WAS OBTAINED FROM CHR. HANSENS LABORATORY. S. CREMORIS PD-1 AND S. LACTIS SUBSP. DIACETYLACTIS CH-1 WERE ISOLATED FROM THE COMMERCIAL MIXED STRAIN STARTERS PD-4 AND CHXL, RESPECTIVELY. ALL CULTURES WERE INCUBATED AT 25°C.

RESULTS

DIFFERENT BIOCHEMICAL PARAMETERS WERE ANALYSED IN THE THREE MILKS AFTER INCUBATION WITH THE STARTERS. ALL RESULTS PRESENTED ARE THE AVERAGE OF DUPLICATE TESTS.

ORGANIC ACIDS OCCUR IN MILK AS A RESULT OF THE HYDROLYSIS OF BUTTERFAT, NORMAL BOVINE BIOCHEMICAL METABOLIC PROCESSES OR BACTERIAL GROWTH. IN ORDER TO INVESTIGATE THE PRODUCTION OR BREAKDOWN OF ORGANIC ACIDS BY BACTERIA, THE CONTENT IN THE THREE MILKS MUST BE MEASURED.

TABLE 1 ORGANIC ACIDS AND PH IN HEAT TREATED COW'S MILK, GI AND GII. CONCENTRATIONS IN PPM

	COW'S MILK		GI		GII	
	x̄	S.E.	x̄	S.E.	x̄	S.E.
PH	6.76	0.02	6.70	0.07	6.41	0.06
OROTIC ACID	64.5	3	20.7	1.8	21.9	1.5
CITRIC ACID	1790	150	1300	120	740	24
LACTIC ACID	[illegible]	36	[illegible]	30	[illegible]	30
URIC ACID	37	3.5	24.6	0.7	20.7	1.2
ACETIC ACID	<30		<30		<30	

THE CONTENT OF OROTIC ACID IS CONSIDERABLY LOWER IN GOAT'S MILK COMPARED TO COW'S MILK. CONCENTRATIONS OF OROTIC AND URIC ACID ARE SLIGHTLY HIGHER IN GII THAN IN GI.

RESULTS WILL BE PRESENTED FOR EACH OF THE BACTERIAL STRAINS

S. CREMORIS PD-1

ANALYSES OF PH AND THE CONTENT OF ORGANIC ACIDS AFTER 16 H INCUBATION ARE SHOWN IN TABLE 2.

TABLE 2 ORGANIC ACIDS AND PH IN COW'S MILK, GI AND GII INCUBATED WITH S. CREMORIS PD-1 AFTER 16H AT 25°C. CONCENTRATIONS IN PPM

	COW'S MILK		GI		GII	
	x̄	S.E.	x̄	S.E.	x̄	S.E.
PH	4.57	0.01	4.57	0.06	4.64	0.02
OROTIC ACID	40.8	4	15.0	1.2	15.8	1.8
CITRIC ACID	1736	126	1795	72	675	35
LACTIC ACID	[illegible]	126	[illegible]	159	[illegible]	145
URIC ACID	30.6	1.5	32.8	2	25.9	1.8
ACETIC ACID	60	7.5	60	5.8	<30	

A SLIGHT DECREASE OF OROTIC ACID IS OBSERVED IN ALL THREE MILKS, BUT NO BREAKDOWN OF CITRIC ACID COULD BE DETECTED.

VOLATILE AROMA COMPOUNDS ARE PRESENTED IN FIGURE 2.

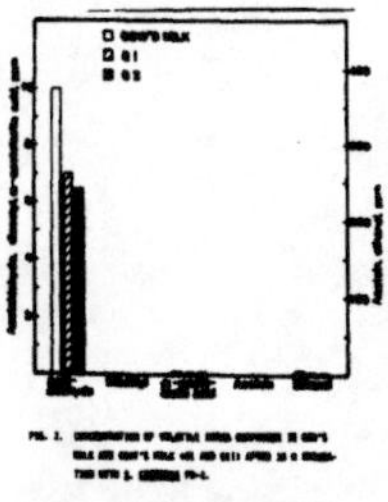

FIG. 2. CONCENTRATION OF VOLATILE AROMA COMPOUNDS IN COW'S MILK AND GOAT'S MILK (GI AND GII) AFTER 16 H INCUBATION WITH S. CREMORIS PD-1.

THIS CULTURE HAD NO BREAKDOWN OF CITRATE, AND CONSEQUENTLY NO PRODUCTION OF DIACETYL, α-ACETOLACTIC ACID OR ACETOIN WAS DETECTED. HIGH PRODUCTION OF ACETALDEHYDE AND LOW CONTENT OF ETHANOL WAS OBSERVED.

S. LACTIS SUBSP. DIACETYLACTIS CH-1

NO CITRIC ACID WAS OBSERVED IN THE GOAT MILKS AND ONLY SMALL AMOUNTS WERE DETECTED IN COW'S MILK AFTER 16 H INCUBATION.

TABLE 3 ORGANIC ACIDS AND PH IN COW'S MILK, GI AND GII INCUBATED WITH S. LACTIS SUBSP. DIACETYLACTIS CH-1 FOR 16 H AT 25°C. CONCENTRATIONS IN PPM

	COW'S MILK		GI		GII	
	x̄	S.E.	x̄	S.E.	x̄	S.E.
PH	4.42	0.02	4.00	0.05	3.71	0.02
OROTIC ACID	40.7	6.5	12.8	0.6	14.8	0.9
CITRIC ACID	130	30	28	12	<5	
LACTIC ACID	[illegible]	150	[illegible]	126	[illegible]	136
URIC ACID	17.0	1.5	11.5	1.7	15.0	1.6
ACETIC ACID	600	52	[illegible]	26	270	36

A SLIGHT DECREASE OF THE OROTIC ACID IS ALSO NOTICED. THE CONTENT OF ACETIC ACID IN THE THREE MILKS CORRESPONDED WITH THE INITIAL CONCENTRATION OF CITRATE.

RESULTS FROM THE GASCHROMATOGRAPHIC ANALYSIS ARE SHOWN IN FIGURE 3.

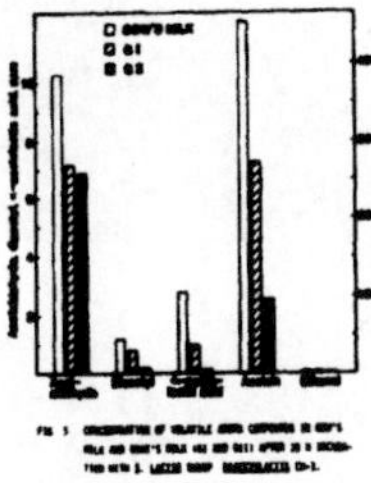

FIG. 3 CONCENTRATION OF VOLATILE AROMA COMPOUNDS IN COW'S MILK AND GOAT'S MILK (GI AND GII) AFTER 16 H INCUBATION WITH S. LACTIS SUBSP. DIACETYLACTIS CH-1.

RATHER LOW LEVELS OF DIACETYL AND α-ACETOLACTIC ACID WAS MEASURED AFTER 16 H, ESPECIALLY IN GII. THE CONTENT OF ACETOIN WAS HIGH.
THIS CULTURE IS CHARACTERIZED BY HIGH PRODUCTION OF ACETALDEHYDE AND LOW LEVELS OF ETHANOL.

LC. CREMORIS CH-1

LEUCONOSTOCS ARE A PARTICULAR FASTIDIOUS GROUP OF ORGANISMS AND GROW POORLY IN MILK UNLESS A STIMULANT IS ADDED. HIGH PH AND LOW CONTENT OF LACTIC ACID AFTER INCUBATION WITH LC. CREMORIS CH-1 IS IN AGREEMENT WITH THIS STATEMENT.

TABLE 4. ORGANIC ACIDS AND PH IN COW'S MILK, GI AND GII INCUBATED WITH LC. CREMORIS CH-1 FOR 16 H AT 25°C. CONCENTRATIONS IN PPM.

	COW'S MILK		GI		GII	
	x̄	S.E.	x̄	S.E.	x̄	S.E.
PH	[illegible]	[illegible]	[illegible]	[illegible]	[illegible]	[illegible]
OROTIC ACID	[illegible]	1.3	[illegible]	1.2	20.7	[illegible]
CITRIC ACID	[illegible]	37	[illegible]	70	[illegible]	40
LACTIC ACID	[illegible]	25	[illegible]	37	[illegible]	35
URIC ACID	20.3	[illegible]	31.3	1.3	20.4	[illegible]
ACETIC ACID	[illegible]	15	[illegible]	15	[illegible]	11

THE RESULTS SHOW CONSIDERABLY BETTER GROWTH OF THIS CULTURE IN GOAT'S MILK THAN IN COW'S MILK. AS LEUCONOSTOCS HAVE LOW PROTEOLYTIC ACTIVITY, BETTER GROWTH OF LC. CREMORIS CH-1 IN GOAT'S MILK COULD BE DUE TO HIGHER CONTENT OF NPN IN THIS MILK. A POSSIBLE ADDITIONAL FACTOR IS STIMULATION BY AMINO ACIDS, OF [illegible].

A SLIGHT BREAKDOWN OF CITRATE AND PRODUCTION OF ACETATE WAS OBSERVED IN ALL MILKS.

ANALYSIS OF VOLATILE AROMA COMPOUNDS REVEALED NO PRODUCTION OF DIACETYL, α-ACETOLACTIC ACID AND ACETOIN. HIGH LEVELS OF ETHANOL WAS OBSERVED, PARTICULARLY IN THE GOAT MILKS.

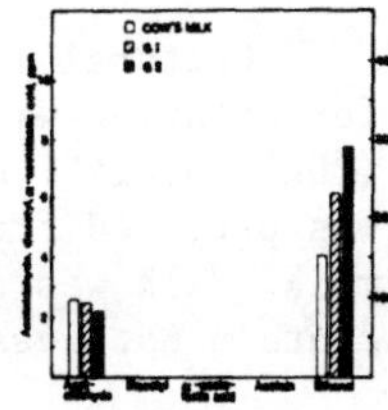

FIG. 4. CONCENTRATION OF VOLATILE AROMA COMPOUNDS IN COW'S MILK AND GOAT'S MILK (GI AND GII) AFTER 16 H INCUBATION WITH LC. CREMORIS CH-1.

LC. CREMORIS CH-1 IN MILK FORTIFIED WITH 0.2% YEAST EXTRACT

ADDITION OF YEAST EXTRACT TO ALL MILKS GAVE SIGNIFICANTLY BETTER GROWTH AND ACID PRODUCTION OF LC. CREMORIS CH-1.

TABLE 5 ORGANIC ACIDS AND PH IN COW'S MILK, GI AND GII INCUBATED WITH LC. CREMORIS CH-1 FOR 16 H AT 25°C. ALL MILKS WERE FORTIFIED WITH 0.2% YEAST EXTRACT. CONCENTRATIONS IN PPM

	COW'S MILK		GI		GII	
	x̄	S.E.	x̄	S.E.	x̄	S.E.
PH	[illegible]	0.03	4.57	0.02	[illegible]	0.01
OROTIC ACID	19.1	4	17.5	2.1	19.8	2.4
CITRIC ACID	111	46	76	15	<5	
LACTIC ACID	7085	400	7587	100	7700	100
URIC ACID	16.7	1.0	13.1	1.2	15.6	1.0
ACETIC ACID	600	75	623	30	405	25

STILL, LOWEST PH WAS OBSERVED IN GOAT'S MILK II AND HIGHEST IN COW'S MILK. A POSSIBLE REASON FOR THIS IS THAT GOAT'S MILK CONTAINS GROWTH FACTORS WHICH LACKS COMPLETELY OR THE LEVEL IS TOO LOW IN YEAST EXTRACT. ALMOST COMPLETE BREAKDOWN OF CITRATE AND GREAT PRODUCTION OF ACETATE WAS REGISTERED IN ALL MILKS.

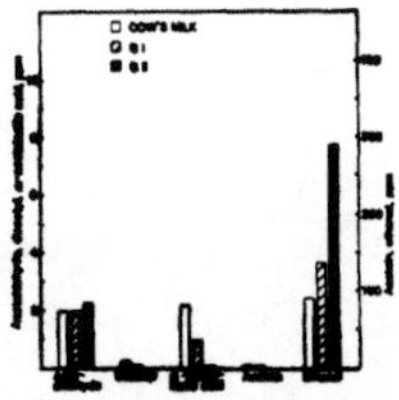

FIG. 5. CONCENTRATION OF VOLATILE AROMA COMPOUNDS IN COW'S MILK AND GOAT'S MILK (GI AND GII) AFTER 16 H INCUBATION WITH LC. CREMORIS CH-1. ALL MILKS FORTIFIED WITH 0.2% YEAST EXTRACT

GOAT'S MILK II HAD ONLY TRACES OF α-ACETOLACTIC ACID AND NO DIACETYL OR ACETOIN WAS DETECTED AFTER 16H INCUBATION. THESE COMPOUNDS WERE FOUND IN LOW LEVELS IN THE TWO OTHER MILKS. THE PRODUCTION OF ETHANOL, HOWEVER, WAS HIGH, PARTICULARLY IN GII.

LITERATURE

RYSSTAD, G. AND R.K. ABRAHAMSEN, 1983
"FERMENTATION OF GOAT'S MILK BY TWO DL-TYPE MIXED STRAIN STARTERS" JOURNAL OF DAIRY RESEARCH 50 349-356

CULTURAL MILK CONTAINING VIABLE BIFIDOBACTERIA

A.V. Gudkov, T.M. Ervolder, M.Ja. Gudkova
All-Union Research Institute of Butter and Cheesemaking
152620, Uglich, Krasnoarmeiskii Boulevard, 19, USSR

Cultural milks ("Uglichskii" and "Vita") were obtained by the fermentation with Bifidobacterium bifidum and lactic streptococci ("Uglichskii") or Lactobacillus acidophilus ("Vita"). Bulk starter was prepared from the freeze-dried concentrate with one transfer in milk. Milk was clotted by 5% of bulk starter for 5 - 7 h. Milks contain not less than 10° cfu/ml of B.bifidum.

KOUWANLAO----TRADITIONAL ROYAL COURT MILK JUNKET OF CHINA

BY JIN SHI-LIN

Inner Mongolia Light Industry Scientific
Research Institute, 30 East Hairlar Road
Huhhot Inner Mogolia P.R.China

SUMMARY. A kind of traditional China Royal Court milk junket is made from cow's milk with press extract of fermented glutinous rice of high-grade SHAOXING rice wine.The milk is coagulated by Mucor and Rhizopus which contained in the extract of fermented glutinous rice of SHAOXING wine.The texture of milk curd is homogenious,tender and delicate,no any whey separated.It has a distinctive flavour with some SHAOXING wine taste.

A kind of traditional China Royal Court milk junket is made from cow's milk with press extract of fermented glutinous rice of high-grade SHAOXING rice wine.The milk is coagulated by Mucor and Rhizopus which contained in the extract of fermented glutinous rice of SHAOXING wine. The texture of milk curd is homogenious,tender and delicate no any whey seperated. It has a distinctive flavour with some SHAOXING wine taste.

Preparation of press extract of fermented glutinous rice Raw materials.

1.Glutinous rice The physical properties of glutinous rice production from the province JIANGSU of China shown in Table 1.

Table 1. Physical properties of glutinous rice

Bulk density		82.30
Size of grain	length (mm)	3.6-6.2
	wide (mm)	2.0-2.8
	thickness (mm)	1.6-2.0
Weight of 1000 grains (g)		20.3

The chemical composition of glutinous rice shown in Table 2.

Table 2. Chemical composition of glutinous rice

	content (%)
Moisture	13.08-15.32
Soluble non nitrogenous compound	68.74-73.13
Protein	5.13-8.13
Crud fiber	0.39-0.62
Ash	0.80-0.99

2. Water Natural water are used,which are no colour,no odour,no taste,clear and transparent.Physical chemical property and composition of water shown in Table 3.

Table 3. Physical chemical property of water

	content (mg/L)
Permanent hardness ($CaCO_3$)	2.0
Temporary hardness ($CaCO_3$)	49.2
Total solids	114
Loss of ignition	55
Free ammonia	0.01
Organic nitrogen total	0.33
Nitrite nitrogen	0.002
Nitrate nitrogen	0.03
Fe	0.35
K	2.5
Na	9.7
Si	7.53
Ca	7.65
Mg	3.81
Phosphate	1.6
Chloride	14.8
Sulphite	10.5

Starter culture

Microorganism from fermented glutinous rice mainly are the Mucor and Rhizopus with some yeast and Absidia.
Mucor javanicus and Mucor Casei are the majority and Rhizopus Chinensis, Rhizopus hangchew, and Rhizopus javanicus also are the majority.

Preparation of the starter

The processing flowsheet is shown in Fig 1. which its distinguishing feature was added a kind of China medicinal herb the powdered polygonum flaccidum.

```
                      grond rice
                          ↓ 20kg
                      ----------
water (11L) →         | Mixing |  ← powdered polygonum flaccidum 150g
                      ----------
                          ↓
                      -----------
                      | Kneading |
                      -----------
                          ↓
                      -----------
                      | Moulding |
                      -----------
                          ↓
                      -----------
Culture (500g) —→     | Inoculum |
                      -----------
                          ↓
                 ----------------------
                 | Culture propagation |
                 ----------------------
                          ↓
                      -----------
                      | Drying   |
                      -----------
```

↓
starter product

Fig.1 The processing flowsheet of preparation of the starter.

Physical character of starter

The starter product's characteristic properties are shown as follow:

Shape------------------------ball type spherular globe
Size dia.(cm)----------------2.0-2.5
Weight per ball(g)-----------5-10
Colour-----------------------milky white
Main component---------------rice powder
additive---------------------polygonum flaccidum powder

Chemical composition of starter

The gross chemical composition of starter is shown in Table 4.

Table 4. Chemical composition of starter (%)

Moisture	13.5-14.9
Crude protein	7.8-8.3
Fat	1.3-1.5
Crude fiber	0.5-0.6
Carbohydrate	66.7-72.1

Microorganisms in starter

Approximate count of microorganisms in the starter are shown in Table 5.

Table 5. Microoganisms in the starter

	approximate count per gram of starter
Mucor	2000
Rhizopus	2000
Absidia	100,000
Aspergillus	2000
Yeast	a great many

Biochemical properties of starter

Some biochemical properties of the starter are shown in Table 6.

Table 6. Biochemical properties of starter

Liquifying ability(mg starch/g starter)	94
Saccharifying power(mg dextrose/g starter)	84
Fermenting power(of the fermenting power as 100 while 1.75g CO_2 production per g starter)	16
Yeast content(per g starter)	16×10^5

Preparation of the press extract

Processing flowsheet of the production of the press extract from the fermented glutinous rice are shown in Fig.2.

```
                         glutinous rice
                               ↓
                          -----------
                          | Sifting |
                          -----------
                               ↓
                          -----------
                          | Dipping |
                          -----------
                               ↓
                          -----------
                          | Cooking |
                          -----------
                               ↓
                    --------------------
water →             | Shower cooling   |
                    --------------------
                               ↓
                          -----------
                          | Inoculum |   ← starter
                          -----------
                               ↓
                        ---------------
                        | Fermentation |
                        ---------------
                               ↓
                           --------
                           | Press |
                           --------
                               ↓
                            extract
                               ↓
                         -------------
                         | Heating    |
                         -------------
                               ↓
                         -------------
                         |  Cooling   |
                         -------------
                               ↓
                         -------------
                         | Settling   |
                         -------------
                               ↓
                         -------------
                         | Decantation |
                         -------------
                               ↓
                         Press extract
```

Fig.2. Processing flowsheet of the production of press extract.

Production of KOUWANLAO by press extract of fermented glutinous rice with milk.

Processing flowsheet of the production of KOUWANLAO are shown in Fig. 3.

```
                milk
                 ↓
          ---------------
          | Filteration |
          ---------------
                 ↓
          ---------------
          |Clarification|
          ---------------
                 ↓
         -----------------
         |Pasteurization |
         -----------------
                 ↓
            ----------
            | Cooling|
            ----------
                 ↓
          ---------------
          | Inoculum    |  ←——press extract of fermented
          ---------------      glutinous rice
                 ↓
          ---------------
          |  Stirring   |
          ---------------
                 ↓
 ---------------------------------------------
 |Filling in cups or porcelain little bowls  |
 ---------------------------------------------
                 ↓
          ----------------
          | Propagation  |
          ----------------
                 ↓
          ----------------
          |  Curdling    |
          ----------------
                 ↓
            -----------
            | Cooling |
            -----------
                 ↓
          Product KOUWANLAO
```

Fig.3. Processing diagram of the production of KOUWANLAO.

Composition and property of KOUWANLAO

KOUWANLAO commercial product with composition and property as follows:

Moisture (%)	83.5
Protein (%)	4.6
Fat (%)	2.1
Carbohydrate (%)	9.2
Ash (%)	1.1
Ca (mg%)	150
P (mg%)	140
Fe (mg%)	0.2

Texture:	Homogenious Softcurd,tender and delicate, no whey separated.
Colour:	white.
Flavour and taste:	distinct flavour with some SHAOXING wine taste.

PRODUCTION OF FATTY ACIDS AND PARTIAL GLYCERIDES FROM MILK FAT TRIGLYCERIDES BY IMMOBILIZED CANDIDA CYLINDRECEA LIPASE

M.K. Tahoun
UNARC, P.O. Box 832, 162 Horreya Avenue, El-Shatby, Aléxandria, Egypt.

Flavour and texture of many dairy products can be improved by partial hydrolysis of milk fat triglycerides. Thus *Candida cylindrecea* lipase was immobilized by covalent coupling to large beads and applied for continuous hydrolysis of milk fat triglycerides. The enzyme lipase complex was used 8 times, recovered intact and hydrolyzed 23% of the substrate within 30 min.

THE EFFECT OF UHT TREATMENT ON THE DENATURATION OF WHEY PROTEINS

J.F. HARDHAM and J.G. ZADOW

AIM

The functional properties of whey products are affected by the degree of protein denaturation. This study aims at determining the effect of pH on the denaturation of Cheddar cheese whey when subjected to UHT processing.

METHOD

Cheddar cheese whey samples were adjusted from pH 7.4 to 6.0 in 0.2 unit intervals and UHT treated on a pilot-scale direct steam injection system at 140 C for 3 sec.

The sediment volumes in each sample after UHT processing were found by centrifugation of 40mL aliquots of whey at 200g for 30min.

The protein content of the supernatant samples was determined by a semi-micro Kjeldahl method.

ANALYTICAL RESULTS OF A TYPICAL RUN

pH	Sediment (mL)	% Protein in Supernatant
7.2	0.05	90
7.0	0.1	92
6.8	0.1	88
6.6	0.2	82
6.5	0.5	73
6.3	0.8	56
6.2	0.8	53
6.0	0.8	48

HPLC ANALYSIS

The supernatant samples were analysed by HPLC using a reverse phase column with an acid saline/ acetonitrile gradient. Samples were adjusted to pH 2.1 and filtered prior to injection into the HPLC column.

C.S.I.R.O. DIVISION OF FOOD RESEARCH, DAIRY RESEARCH LABORATORY, MELBOURNE, AUSTRALIA.

The graphs show the effect of pH on denaturation of beta-lactoglobulin and alpha-lactalbumin for 3 separate trials.

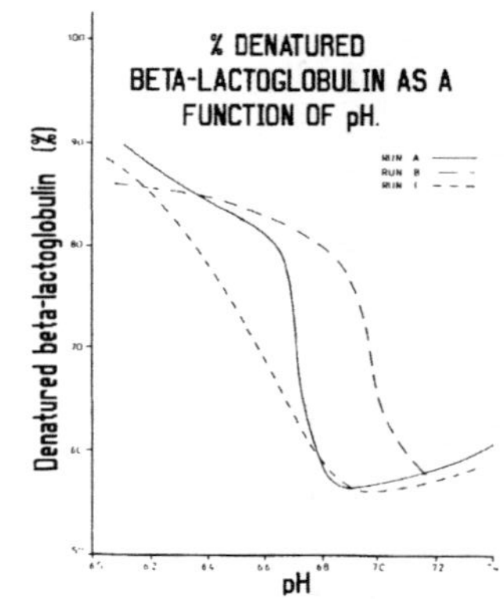

The degree of denaturation of beta-lactoglobulin increased with decreasing pH.

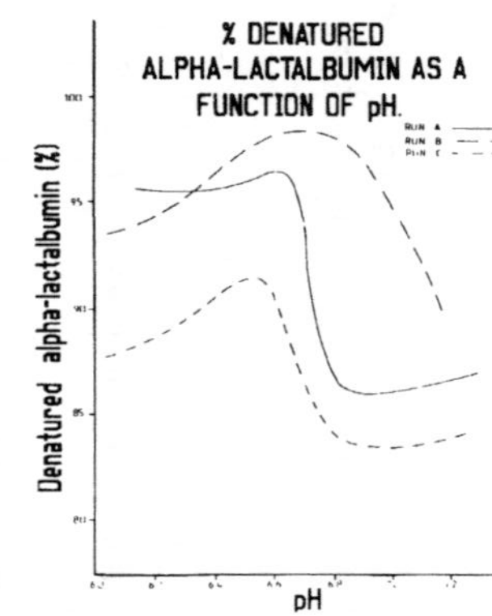

The degree of denaturation of alpha-lactalbumin showed unexpected behavior. Initially, the denaturation increased as the pH decreased from 7.4. There was a reversal in this trend below pH 6.6.

LACTOSE HYDROLYSIS OF DAIRY PRODUCTS

J.F.Hayes, I.Mitchell, M.Free and J.G.Zadow

A commercial process, utilizing lactase immobilized on amphoteric ion-exchange resin (IMS), commenced operation in Australia in June, 1985. This commercialisation is the culmination of research by a group of companies* into column design, cleaning sanitation, enzyme life and product functionality. The major advantage of this process is its ability to hydrolyze feedstocks with a wide range of pH. At a reaction temperature of 40C, the enzyme can process approximately 15, 40 and 60 bed volumes/hr of milk, whey and permeate respectively, with a degree of hydrolysis of greater than 60%.

*GROUP MEMBERS: SUMITOMO, APV/BELL BRYANT, ABC, MILES (AUSTRALIA), CSIRO, HYDROLYSED PRODUCTS.

C.S.I.R.O. DIVISION OF FOOD RESEARCH, DAIRY RESEARCH LABORATORY, MELBOURNE, AUSTRALIA.

Process Flow Chart for Hydrolysis of Feeds

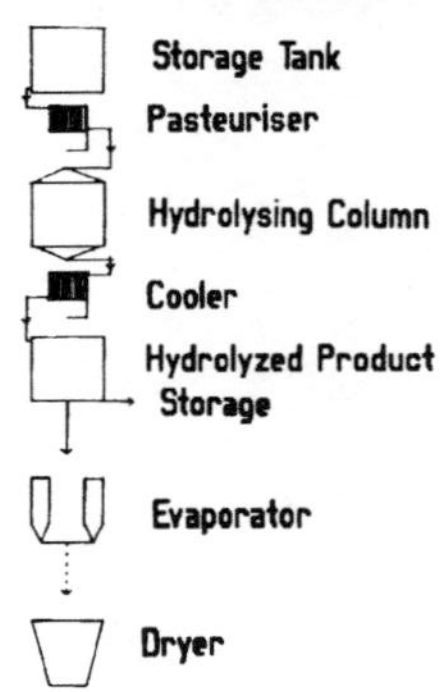

Effect of Temp. on Activity of IMS.

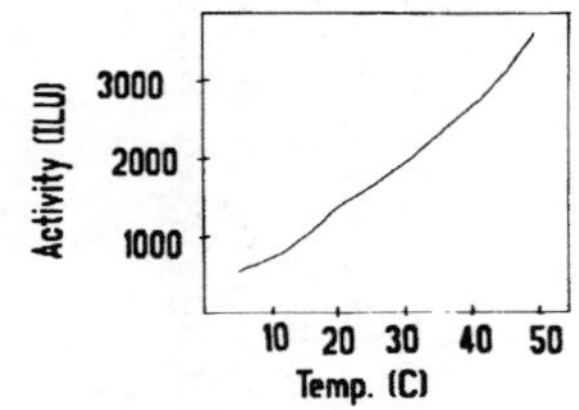

Lowering reaction temp. increases the half-life in column operation.

pH Profile Curve of IMS

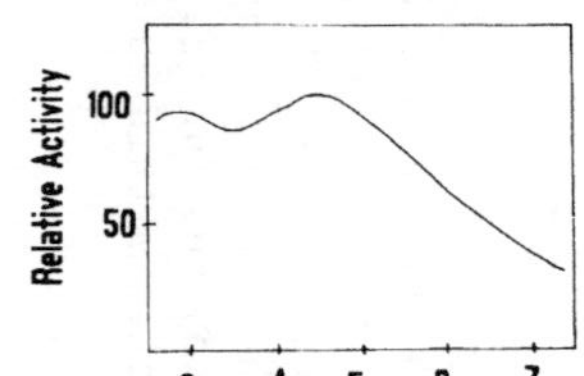

Physical Properties of IMS

Appearance: pinkish yellow granules.

Activity: 1000, 1800, 2000 ILU/g of IMS.

1 ILU (International Lactase Unit) is defined as the amount of enzyme necessary to liberate one micromole of D-glucose in one minute.

IMS is stable in the range of pH 2 to 8 and has operational activity from pH 2.5 to 7.

The optimum activity is between pH 4 and 5.

Benefits of Hydrolyzed Products.

1. Can be utilized by lactose malabsorbers.
2. Sweeter product without additives.

HYDROLYSIS OF MILK FAT TRIGLYCERIDES BY AN IMMOBILIZED LIPASE

M.K. Tahoun, M. Youssef, S. Abou-Donia
UNARC, P.O. Box 832, 163 Horreya Avenue, El-Shatby,
Alexandria, Egypt.

Cheese flavour can be enhanced by partial lipolysis of milk fat. The recent price increase of pancreatin recommends the application of immobilized lipases. Thus pancreatic lipase was immobilized by various methods, however large agarose beads were more convenient as they easily recovered for reuse 8 times and hydrolyzed 31% of milk fat triglycerides within 30 min.

Enzymatic synthesis of new trisaccharide, isoraffinose, from a mixture of lactose and sucrose

Kyozo Suyama and Susumu Adachi
Faculty of Agriculture, Tohoku University,
Tsutsumidori-Amamiyacho 1-1, Sendai, 980 JAPAN

The enzymatic transgalactosylation reaction of lactose in the presence of sucrose as an acceptor was studied. The main product of E.coli β-galactosidase was GS (see Fig. 1) which was a hitherto undescribed trisaccharide, O-β-D-galactopyranosyl-(1⟶6)-O-α-D-glucopyranosyl-(1⟷2)-O-β-D-fructofranoside (FAB/MS m/z:527($\overline{M}$ + Na^+), ^{13}C-NMR (see Fig. 2), hydrolysis with β-fructosidase yielded fructose and allo-lactose(O-β-D-galactopyranosyl-(1⟶6)-D-glucose), to which we assigned the trivial name isoraffinose. Sucrose was an excellent acceptor to accumulate isoraffinose in good yield (see Table 1). Comparative rates of hydrolysis of isoraffinose against lactose and sucrose by β-galacto-sidase and β-fructosidase, respectively, were determined (see Table 2).

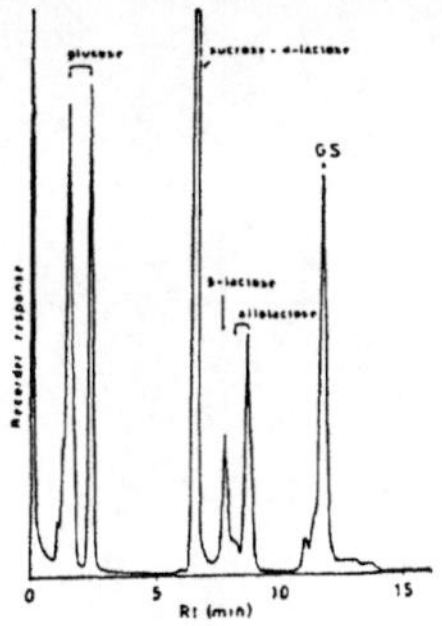

Fig. 1. GC of TMS-derivatives of trans-galactosilation products of lactose-sucrose system

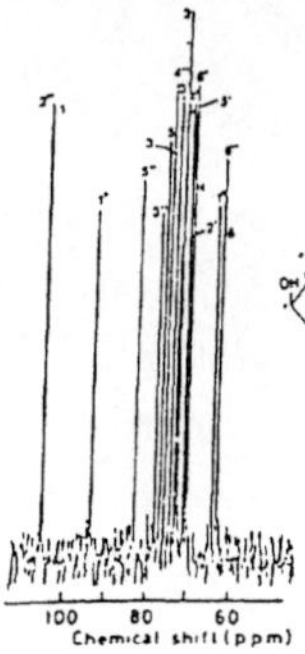

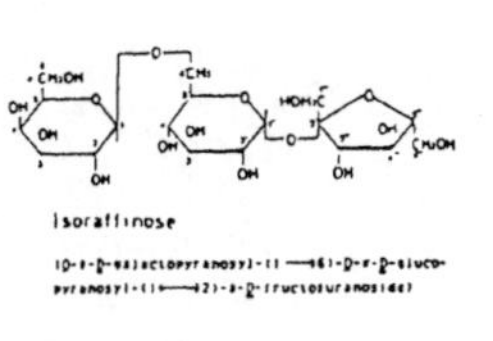

Fig. 2. ^{13}C-NMR of isoraffinose

Table 1 Sugars composition (%) of the enzymatic transgalactosidation reaction products

	5hr*	24hr*
Galactose	-	Trace
Glucose	8.5	22.8
Sucrose	45.9	39.4
Lactose	35.8	8.1
Allolactose	4.4	8.9
Isoraffinose	4.9	17.0
Others	0.3	2.5

* Incubation time

Table 2. Relative rates of hydrolysis of isoraffinose by enzymes.

Source of enzyme	isoraffinose	lactose	sucrose
β-galactosidase			
Asp. orizae	65.5	100	
E. coli	28.7	100	
Klu. lactis	2.1	100	
β-fructosidase			
yeast	22.2		100

SURFACE TENSION OF WHEY PROTEINS STUDIED BY A DROP-VOLUME APPARATUS

M. Paulsson
Department of Food Technology, University of Lund,
P.O. Box 124, S-221 00 Lund, Sweden

Proteins are amphiphilic macromolecules which make them suitable surface-active agents in food emulsions. The surface tension of whey proteins was studied according to the drop-volume method as a function of time at different concentrations and pH-values. Interactions between proteins, lipids and carbohydrates in milk have also been studied by this method.

INTER-ESTERIFICATION OF BUTTER FAT SOLID FRACTION/RAPESEED OIL MIXTURES WITH CANDIDA CYLINDRACAE LIPASE AS CATALYST

P. Kalo, K. Vaara* and M. Antila*
Departments of General Chemistry and *Dairy Science
University of Helsinki
SF-00710 Helsinki
Finland

ABSTRACT. Butter fat solid fraction/rapeseed oil mixtures were inter-esterified with Candida cylindracae lipase as catalyst. The changes in triacylglycerol composition induced by inter-esterification were followed by quantitative determination of triacylglycerols with different acyl carbon number and level of unsaturation by high resolution gas chromatography. The melting and crystallisation properties of the reaction products were determined by DSC method.

1. INTRODUCTION

The alteration of the physical properties of fats through inter-esterification is currently being studied as a mean of making them more suitable for the manufacture of shortenings, margarines and confectionary fats. The inter-esterification reactions catalysed by immobilised lipase enzymes have been treated in a number of recent reports. Inter-esterification with nonspecific lipase as catalyst yields a fat similar to that obtained with chemical catalysts, where the fatty acids are randomly distributed among the triacylglycerols (1,2). Here we report on inter-esterification reactions of butter fat solid fraction/rapeseed oil mixtures catalysed by Candida cylindracae lipase . The changes induced in the triacylglycerol composition were followed by determination of the triacylglycerols of different acyl carbon numbers and levels of unsaturation by high resolution gas chromatography, and the changes in the melting and crystallisation properties were followed by DSC method.

2. MATERIAL AND METHODS

Butter fat solid fraction S_{24}/rapeseed oil (70/30) and butter fat solid fraction S_{24}/hydrogenated rapeseed oil (70/30) mixtures dissolved in hexane were heated with Candida cylindracae lipase on Celite for 14 days at 35 ^{o}C with water as activator.

The triacylglycerols of different acyl carbon number and levels of unsaturation were determined on a 10-m immobilised SE-54 silica capillary column with a 0.10 μm film thickness and 0.2 mm I.D (3), and the melting and crystallisation curves were recorded with a Perkin-Elmer DSC-4.

The random triacylglycerol compositions of the fat mixtures were calculated by microcomputer according to the equations

$$\% \text{ AAA} = A^3/10000$$
$$\% \text{ AAB} = 3A^2B/10000$$
$$\% \text{ ABC} = 6ABC/10000,$$

where A, B and C are the concentrations of fatty acids expressed in mol per cent and AAA, AAB and ABC are triacylglycerols composed of one, two and three fatty acids, respectively.

3. RESULTS

In both reaction mixtures, the measured values for proportions of trisaturated triacylglycerols were for most peaks nearly the same as the calculated values. The proportions of monoene triacylglycerols in the inter-esterified fat mixtures were likewise on about the same level as in the calculated mixtures. Diacylglycerols present in the reaction products caused the differences in the acyl carbon number range 30-38. Their approximate proportions were determined.

The melting curves of untreated mixtures and reaction products show that the melting range was narrower, the proportion of fat melting at high temperature range lower and the proportion of fat melting at 0-20 °C higher in the reaction products.

4. CONCLUSIONS

In both mixtures, the greatest changes in triacylglycerol composition induced by inter-esterification reaction were the decrease of trisaturated triacylglycerols with 42-52 acyl carbons and the increase of monoene triacylglycerols with 48-52 acyl carbons.

The melting curves of the reaction products showed narrower melting ranges than the untreated mixtures, and lower proportions of fat melting in the high temperature range.

5. REFERENCES

1. A.R. Macrae, J. Amer. Oil Chemists' Soc. **60** 291 (1983).
2. P. Kalo, P. Parviainen, K. Vaara, S. Ali-Yrkkö and M. Antila, Milchwissenschaft **41** 82 (1986).
3. P. Kalo, K. Vaara and M. Antila. J. Chromatogr., in press.

VALORIZATION OF PHOSPHOLIPIDS FROM DAIRY BY-PRODUCTS

Y. Van Coillie and A. Huyghebaert
Laboratory of Food Chemistry and -microbiology
Faculty of Agricultural Sciences
Coupure links, 653 B-9000 GHENT Belgium

Phospholipids are good emulsifiers and commercial preparations are extensively used in manufactured foods. Butter milk, butter serum, skimmed and aqueous phases from the production of anhydrous milk fat are important sources of phospholipids. The purpose of this work is to investigate different separation techniques with the aim of isolating phospholipid rich fractions.
A scheme for the enrichment of phospholipids from butter milk and butter serum is presented.

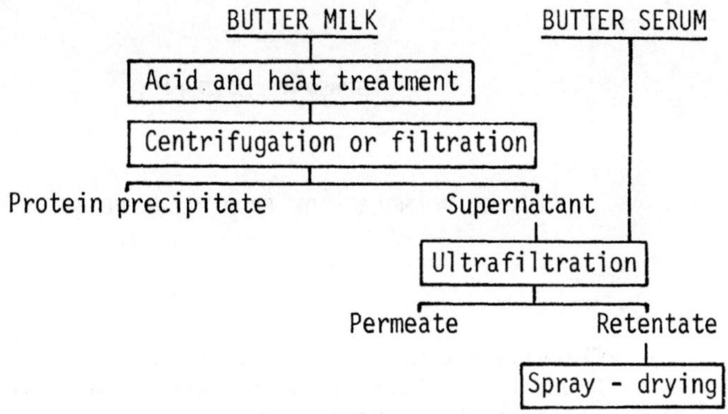

Ultrafiltration is not an appropriate method for the isolation of phospholipids from skimmed and aqueous AMF phases due to the low permeability of the proteins present
The valorization of phospholipids from butter milk is possible by a casein precipitation and a concentration by an ultrafiltration technique.
Phospholipids from butter serum can be enriched by ultrafiltration, without a prior protein precipitation.
A critical point is the sensitivity to oxidation. Flushing with nitrogen or addition of anti-oxidants is necessary to protect phospholipids against degradation reactions.

WHEY PROTEIN CONCENTRATES AS BINDERS IN SALT FREE RESTRUCTURED MEATS

A.M. Fayerman* and R.F. Mawson**
* New Zealand Dairy Research Institute, Private Bag, Palmerston North, New Zealand.
**Meat Industry Research Institute of New Zealand Inc., P.O. Box 417, Hamilton, New Zealand

The salt and polyphosphate solubilized muscle protein, myosin, is traditionally used as the binding agent in the manufacture of restructured meats. Products so prepared however suffer from salt and phosphate induced oxidation upon storage. Gelation-enhanced whey portein concentrates (WPCS) were used to successfully replace solubilized myosin as the binder in restructured meats. Their use retarded storage related oxidation defects associated with salt and phosphate.

THURSDAY – POSTER 40

INFANT FORMULA MADE FROM MILK AND SOYBEAN

Changjiang Liu, Changyu Ai, Yunchun Du
Heilongjiang Dairy Industry Research Institute,
113 Xuefu Road, Harbin, P.R. China

This product suitable for infant feeding contains cow's milk solid, soybean extraction solid and additional carbonhydrates. The ratio of those is 6:1:3. Vitamins and minerals are also added. Infant-feeding experiment indicated better result comparing to sweetened whole milk powder as a substitute for breast milk.

THE APPLICATION OF CONDUCTIVITY IN THE MILK QUALITY TESTING IN CZECHOSLOVAKIA

R. Janál, J. Svasta, J. Louds*
Univerisy of Agriculture, Prague /*Research and Development Institute of the Tractor Station and District Agricultural Administration, Czechoslovakia

The conductivity of milk has been used as a measure of milk quality testing both in Czechoslovak large-scale dairy farms and in the central milking parlours. A number of conductivity measurement instruments were designed, some of them being now manufactured; on the farm, and or under quarter milk examinations.

INTERLABORATORY COMPARISON PROGRAMMES IN THE NEW ZEALAND DAIRY INDUSTRY: CURRENT STATUS

R.C. LEONG, M.J. DYKES AND P.J. SPARKSMAN
MINISTRY OF AGRICULTURE AND FISHERIES
NATIONAL DAIRY LABORATORY
RUAKURA AGRICULTURAL RESEARCH CENTRE, HAMILTON NEW ZEALAND

The New Zealand dairy industry exports about 80% of its milk as dairy products. Therefore, certification of the chemical and microbiological content of these products by an independent body (Ministry of Agriculture and Fisheres, MAF) to ensure the safety and integrity plays an important role to our export drive. One of the ways MAF controls the competency of industry laboratories is to require them to participate in a comprehensive interlaboratory comparison programme (ILCP).

Since 1977, MAF has conducted ILCP covering over 60 dairy laboratories. The ILCP handles 9 different dairy products and the number of tests performed by all these laboratories averages about 50,000 per annum.

Before a nationwide ILCP can operate effectively, certain criterion must be met. These are:

(a) Participating laboratories are to use standard method of analysis.

(b) Participating laboratories are to treat ILCP samples as routine samples.

(c) Criteria for non-acceptable performance on tests must be established.

(d) Use of control or reference samples are encouraged to assist testing conditions.

(e) A national body present to co-ordinate the feedback of results rapidly to users.

(f) A body to review data generated and the needs of the industry.

The ILCP plays an important role in the monitor of laboratories performance in the New Zealand dairy industry. The programme is also used as a "quality control tool" by participating laboratories in the day to day running of their laboratories.

MONITORING OF FAT CRYSTALLISATION

B. Grönlund, M. Heikonen, T. Moisio
Valio Finnish Co-operative Dairies' Association
Research and Development Department
P.O. Box 176
SF-00181 Helsinki
Finland

Anhydrous butter fat was crystallised on a laboratory scale. The crystalliser consisted of a stirred glass vessel in a micro-processor controlled cooling bath. A slow paddle agitator was used to enhance heat transfer in order to eliminate temperature gradients which would cause heterogenous subcooling.

The molten fat was crystallised by cooling of the bath according to preset stepwise linear temperature profiles. The thermal and rheological behaviour of the melt was monitored through three parameters:

- the temperature of the melt
- the temperature difference between melt and coolant
- the dynamic viscosity of the melt

The two former were measured with a Ni10-sensor in the melt and the Pt100-sensor of the cooling bath thermostat. The latter was measured with a Gelograph oscillation viscosimeter, originally designed for milk coagulation studies. The analog parameter data were digitalised and computer processed.

In addition to the traditional plotting of parameter vs. time an improvement from a process monitoring point of view was made, namely:

- time derivate curves of the parameters.

The derivate curves were computed on-line. The exponential scale of viscosity causes accordingly increasing noise in the derivate curve. Therefore the derivate of the viscosimeter output voltage which is directly proportional to the logarithm of the viscosity was monitored instead.

It was observed that the derivate curves more distinctly evaluated the increase in viscosity, as well as the subsequent temperature rise and the increased temperature difference between melt and coolant caused by the crystallisation enthalpy. The initiation, culmination and termination points of crystallisation were thus clearly revealed. This could be utilized in studying effects of different raw materials and changing process variables such as cooling and stirring rates on the crystallisation of fats.

GPSR Compliance

The European Union's (EU) General Product Safety Regulation (GPSR) is a set of rules that requires consumer products to be safe and our obligations to ensure this.

If you have any concerns about our products, you can contact us on ProductSafety@springernature.com

In case Publisher is established outside the EU, the EU authorized representative is:

Springer Nature Customer Service Center GmbH
Europaplatz 3
69115 Heidelberg, Germany

Batch number: 10620141

Printed by Printforce, the Netherlands